THE ORYX
RESOURCE GUIDE
TO
EL NIÑO AND LA NIÑA

Joseph S. D'Aleo
with Pamela G. Grube

ORYX PRESS

Westport, Connecticut • London

The rare Arabian Oryx is believed to have inspired the myth of the unicorn. This desert antelope became virtually extinct in the early 1960s. At that time, several groups of international conservationists arranged to have nine animals sent to the Phoenix Zoo to be the nucleus of a captive breeding herd. Today, the Oryx population is over 1,000, and over 500 have been returned to the Middle East.

Library of Congress Cataloging-in-Publication Data

D'Aleo, Joseph S.
  The Oryx resource guide to El Niño and La Niña / by Joseph S. D'Aleo with Pamela G. Grube.
    p. cm.
  Includes bibliographical references and index.
  ISBN 1–57356–378–1 (alk. paper)
    1. El Niño Current.   2. La Niña Current.   3. El Niño Current—Environmental aspects.   4. La Niña Current—Environmental aspects.
  I. Grube, Pamela G.   II. Title.
  GC296.8.E4D35   2002
  551.6—dc21   2001050030

British Library Cataloguing in Publication Data is available.

Library of Congress Catalog Card Number: 2001050030
ISBN: 1–57356–378–1

First published in 2002

Oryx Press, 88 Post Road West, Westport, CT 06881
An imprint of Greenwood Publishing Group, Inc.
www.oryxpress.com

Printed in the United States of America

∞™

The paper used in this book complies with the Permanent Paper Standard issued by the National Information Standards Organization (Z39.48–1984).

10 9 8 7 6 5 4 3 2 1

Contents

Color plates follow page 88.

Preface

As I addressed the nation's television weather broadcasters for the third consecutive year on the status and the seasonal effects of the Tropical Pacific phenomena known as El Niño and La Niña, I realized how much has changed in the last few decades.

As late as 1980, when I taught weather and climate to college students, the El Niño phenomenon was discussed only in reference to the weather in the tropical Pacific, from Peru to Australia and Southeast Asia. At that time, the relationship between the sea surface temperatures and the pressure, temperature, and precipitation patterns in the Tropical Pacific was well established and accepted. Some of the more progressive governments in these areas had even begun to monitor these events and to apply the knowledge for decisions and policies related to agriculture and foreign trade. But for the rest of the world, it simply became a story relegated to the science magazines and a paragraph buried in the Sunday paper.

Then came the Super El Niño of 1982 and 1983, which coincided with great climate and weather extremes on a global basis. This spurred considerable media attention, which fostered much government-funded research. This research has lead now to a much greater understanding of the phenomenon and a recognition of its true global importance.

Thanks to this heavy media coverage, the phenomenon, which for centuries was known only by Peruvian fishermen and farmers and climatologists, today has become a household word. Although undeserving of all the blame it gets for virtually every incident of un-expected and severe weather during the events, it does have a major effect on the weather and seasons, no matter where one lives in the world.

WEATHER FORECASTS

There have been more advances in the field of meteorology in the last century than in all of prior recorded history. In the early part of the century, the primary advances were in the understanding of the basic elements and processes of the atmospheric system, including air masses and fronts. It was in the first half century that we saw the first weather "forecasts."

Most of the great advances in meteorology in the last half of the last century have come from the development and application of new technology and from technology-supported research that transformed weather forecasting from an "art" to a science.

A full national network of Doppler radars helps not only track se-vere local storms but also detect rotation within storm clouds ena-bling advanced warning of tornadoes. Polar orbiting and geostationary satellites help monitor global weather and provide early detection of tropical storms. Powerful computers run computer fore-cast models that can resolve differences in forecast conditions down to the very local level or look weeks into the future. There are better sensors on board aircraft, on ships, on buoys, and on thousands of land stations that measure conditions automatically and transmit them back to world meteorological centers for use in these computer mod-els. And, the more data that are input into the models, the more accurate and detailed the output. The total investment in these new technologies runs in the many billions of dollars.

Some other advances have come not from investment in technol-ogy but from investment in weather and climate research. The out-come of research is often an improved understanding of the earth and its atmosphere. Researchers reap the benefits of this investment when they can use this new understanding to provide better forecasts.

One of the best examples of this is the story of the El Niño and its associated Southern Oscillation. Once thought to be phenomena solely of the Tropical Pacific, research, spurred on by the worldwide extremes associated with the Great El Niño of 1982 and 1983, has uncovered a true global connection between this phenomenon with climate anomalies and extremes. In recent years, there has been great success in use of this new knowledge to predict seasonal weather in advance to the benefit of governments, industry, and the people they serve.

The success of this research has encouraged further research into possible climate connections with other oceanic and atmospheric oscillations and phenomena. This promises, some day, to make seasonal forecasts as routine and accurate as the 5-day forecast one views each day on a television weather show, newspaper, or Web site.

THIS BOOK

This book is designed to provide a basic understanding of the two Tropical Pacific siblings now known as El Niño and La Niña. This volume presents the basic causes and effects on weather, the environment, and the global economies and what governments and industries worldwide are currently doing about it.

Chapter 1 covers the history of the El Niño Southern Oscillation (ENSO). Although there was a growing awareness of oscillatory phenomena in both the atmosphere and ocean more than a century ago, recognition that the ocean and atmosphere flip-flops were related did not come until the 1950s. It wasn't until the 1980s that scientists recognized that the phenomena influenced global weather patterns.

Chapter 2 defines the causes of the oscillations and how the atmosphere and ocean work together. Chapter 3 presents the effects of ENSO on the global weather. Effects on weather in the United States are covered in chapter 4, for winter and spring, and in chapter 5, for summer and fall.

Chapter 6 discusses the effects of these phenomena on the ecosystems of the world. This includes human health and the spread of infectious disease and the wide-ranging effects on marine and animal life. Chapter 7 shows how scientists study and track the ENSO phenomena of El Niño and La Niña, including how real-time observations are made and current measures of strength.

Chapter 8 presents advances scientists have made in predicting the ENSO state as much as 12 to 18 months in advance. Because the likely effects of each ENSO state is now known, the ability to forecast has enormous value as it allows forecasters to predict the climate more accurately.

In chapter 9, two Super El Niño events are discussed. The first, the event in 1982–1983, led to the heavily funded research that has provided the exciting new level of understanding and ability to forecast; and the second, the event of 1997–1998, which caused floods, tornadoes, snow in unusual locales, and drought, brought even more understanding. Chapter 10 shows how governments and industries around the world are using this new information, and, despite being known for negative consequences, demonstrates that El Niño and La Niña both bring economic benefits.

Chapter 11 discusses the fact that the ENSO relationship with weather may be just the tip of the iceberg in climate prediction. In the flurry to understand ENSO, scientists have discovered some other relationships that may further extend our ability to forecast seasonal weather.

Chapter 12 offers brief biographies of nine pioneers and key people involved in ENSO research over the past century. Chapter 13 lists the most important organizations involved in ENSO research and forecasting.

At the end of the book, a chronology presents key events in ENSO from the 15th century through the present. A glossary provides basic definitions of frequently used terms in ENSO and climate forecasting. The bibliography gives the reader numerous sources for further reading and research about El Niño and La Niña. Further Reading: The Basic List provides an annotated list of publications that offer the reader a good general understanding of ENSO; Further Reading: A List by Category gives readers, especially those with a more advanced scientific background, a list of nearly 800 sources, ranging from scholarly journal articles to articles in popular magazines and newspapers, and from articles on Web sites and in monographs to monographs and books covering the subject.

Chapter 1

El Niño and La Niña: Children of the Tropics

As recently as the very early 1980s, the term *El Niño* was familiar only to climatologists and readers of scientific journals. Today it attracts widespread media attention and public interest. El Niño now makes one think of winter storms and floods, unusual global warmth, and major third-world droughts and famines. Stark images are presented on television and in the print media of the global catastrophes El Niño has wrought. And scientists continue to provide the historical evidence to show the effects are real and consistent.

Ironically, El Niño is just one of two modes of a naturally occurring climate phenomenon, involving both the atmosphere and ocean across the Tropical Pacific. Although not really discovered until well into the last century, the phenomenon has been going on probably as long as man has walked this great planet.

The phenomenon known as El Niño Southern Oscillation (ENSO) is an interannual, coupled oscillation in the atmosphere and ocean of the Tropical Pacific. In the atmosphere, it is an east–west see-saw of surface pressure and the related patterns of clouds, winds, temperatures, and precipitation. In the ocean, it is an east–west flip-flop of the location and depth of warm and cool pools of water.

Saying that ENSO is a *coupled oscillation* means that the atmosphere and ocean influence each other in a very continuous, cyclical way (Wallace and Vogel, 1994). The atmospheric changes help produce the changes observed in the ocean, and these changes, in turn, act to influence and ultimately change the atmosphere above, which, in turn, bring about further changes in the ocean, and so on. Within the ocean in a macro sense, the ENSO phenomenon manifests itself as a slow sloshing of water in the huge bathtub we call the Pacific Basin. The warm water sloshes to the east and then a year or more later, back to the west. The sloshing is manifested by a series of very large-scale internal waves called Kelvin and Rossby waves.

In the atmosphere, a similar transfer of air takes place that results in a flip-flop in pressure patterns and winds. Although the oscillation is most evident and consistent in the Tropical Pacific, research has shown that global weather patterns are also reliably influenced; and that the effects, like the modes of the oscillation in the Pacific, are opposite from one mode to the other.

The two modes have been given the names El Niño (the Infant Boy) and La Niña (the Infant Girl). Although La Niña, El Niño's lesser known sibling, gets much less media attention, it may actually have more negative effects than El Niño on the climate, at least in the United States. La Niña brings more frequent destructive hurricanes, severe thunderstorms and tornadoes, and costly winter cold waves.

The newly discovered relationships, and the fact that once they begin each mode typically lasts a year or more, has enabled a whole new "science" of seasonal climate forecasting.

BASIC TERMS DEFINED

This section begins with a quick "definition" of the terms used frequently in this book. A more complete glossary is found in the back of the book. Each term is explained further in this book. The causes of the ocean and atmospheric phenomena are described in chapter 2.

The *Southern Oscillation* is an interannual see-saw in sea-level pressure across the Tropical Pacific closely linked with El Niño and La Niña. It was first documented and named by Sir Gilbert Walker in the 1920s (Walker, 1928). He noted there was an inverse relationship between surface air pressures in the eastern and western Tropical Pacific. Higher than normal sea-level pressure in the western Tropical Pacific is almost always concurrent with low pressure in eastern areas and vice versa. It represents a standing wave or "see-saw," a mass of air oscillating back and forth in the Tropical and Sub-Tropical Pacific.

Southern Oscillation Index (SOI) is a measure of the state of the Southern Oscillation first proposed by Walker. It measures the pressure difference between Darwin, Australia and the South Pacific Island of Tahiti. In the SOI, the pressures at both locations are normalized (evaluated relative to normal) to account for the normal variability (e.g., seasonal) in pressure between Tahiti and Darwin.

ENSO is an acronym that stands for El Niño Southern Oscillation. The term is used to describe the full range of the Southern Oscillation including both the warm (El Niño) and cold (La Niña) events. Technically, it includes both the oceanic (warm or cool water) component and the atmospheric (Southern Oscillation) component of the phenomena.

El Niño was a term originally used by 19th-century fishermen to describe a southward moving current of warmer water off the coast of Peru and Ecuador that occurred each year around or shortly after Christmas. The term *El Niño* in Spanish means "the Infant Boy," referring to "the Christ Child." In some years, the warming is more extensive and can last 1 or 2 years. Today, the term *El Niño* is used to describe these large-scale, long-lasting "warm" events and the accompanying large-scale atmospheric changes.

During El Niños, unusually high pressure typically develops in the western Tropical Pacific, while pressures in the eastern Tropical Pacific become unusually low (a negative SOI). This causes a weakening in the trade winds, which can reduce the cool water upwelling. This leads to a subsequent warming of the waters in the eastern Tropical Pacific. These changes in pressures, winds, and water temperatures cause the tropical showers that normally are found in the western Tropical Pacific to shift east to the east central Pacific and in extreme events, all the way to the coast of Peru.

La Niña (in Spanish it means "the Infant Girl") is in many respects the opposite of El Niño. It has also been called El Viejo ("The Old One") and the Anti-El Niño, but this term has fallen into disfavor, as it would imply an "Anti-Christ." This cool relative of El Niño is characterized by a large-scale strengthening of the trade winds and a cooling of the water in the eastern and central Tropical Pacific, due to an upwelling of cool water from beneath. Like El Niños, La Niñas often begin in the summer of the southern hemisphere (December to February) and may last 1 or 2 years.

During La Niñas, pressures rise in the eastern Tropical Pacific while they fall in the western Tropical Pacific (a positive SOI). This causes an increase in the trade winds, enhanced upwelling, and a cooling of the waters in the eastern Tropical Pacific. These changes cause a suppression of showers in the eastern Tropical Pacific and a further enhancement of the normal rainfall in the western Tropical Pacific.

DISCOVERY OF THE EL NIÑO SOUTHERN OSCILLATION

Sailors and farmers were the first weather forecasters. Their livelihoods and safety depended on their ability to recognize potential weather problems from observing and interpreting the sky. It is not surprising that the first to recognize the existence of the recurring phenomenon known today as El Niño were the farmers and fishermen of Peru.

Peruvian Fishermen

Centuries before it was first "discovered" and studied, the fishermen and farmers of Peru were aware of and used this cyclical phenomenon. Most of the time, the waters off the Peru coast are cool and nutrient rich, hosting one of the world's most productive fisheries. The northward moving Peru Current is responsible, as it draws the surface water away from the coast, inducing cool water to well up from the depths.

However, each year, a warmer southward moving ocean current typically appears around Christmas time and lasts for several months. As indicated earlier, the term *El Niño* was first used by 16th-century Peruvian fishermen to refer to this warm ocean current. They chose the name because of the tendency for it to appear around the time of the Christ Child's celebrated birth. Fish tend to be less abundant in the warmer, lower salinity water, therefore the fishermen typically return to port and use this time to repair their boats and nets and spend more time with their families. The warming usually gets no further south than northern Peru and usually ends by March or April.

In some years however, the warming becomes far more extensive and significant, lasting a year or two and leaving a devastating effect on the fisheries. Over recent years, the term *El Niño* has come to refer to these exceptional, long-lasting, and more damaging warm events.

South American Farmers

Potato farmers in the mountains of Peru and Bolivia also may have been able to predict El Niño and the associated drought for centuries by observing the night sky in the months before the growing season. University of California researchers (Orlove, Chiang, & Cave, 2000) found that farmers as far back as the Inca civilization in the 15th century had noted that when all the 11 main stars in the constellation Pleiades were bright in June, abundant rains followed from October to May. When the dimmest five stars were obscured, drought usually followed. Farmers often then delayed planting the potato crop, which is very drought sensitive.

Orlove observed that during El Niños, high-level cirrus clouds are more prevalent over the Andes. The clouds are generated by distant thunderstorms, which develop in relation to the warming in the eastern Tropical Pacific. These clouds dim the stars. El Niños usually produce below normal precipitation in these mountainous regions through the following southern hemisphere growing season.

Today, it is known that these local effects are closely coupled with large-scale atmospheric changes, and further that these events have significant effects on the global climate. However, it took many decades to reach this level of understanding.

EARLY SCIENTIFIC WORK

Most of the early scientific work was focused on the atmospheric changes associated with the phenomenon. It was inspired not by the warming of the waters and disruption of the fishing off of Peru, but by the periodic failures of the monsoon rains on the other side of the Pacific, in places like India. Many of the early researchers were trying to find a cause for the year-to-year monsoon variability.

In fact, it was the worst famine in India's history, caused by the failure of the monsoon in 1877 that really triggered the first intensive research work. An observatory was founded in India to explore the cause and whether the famines could be prevented.

The first to postulate the existence of a large-scale atmospheric oscillation in the Tropical Pacific was H. H. Hildebrandsson in 1897, who discovered it while examining 10 years of surface pressure data. This apparent see-saw in pressure between Indonesia and South America was later reaffirmed in 1902 by Norman and W. J. S. Lockyer with more extensive data.

In 1904, Sir Gilbert Walker, a British mathematician, entered the British Colonial Service as the director general of the observatory, with the goal of predicting the fluctuations of the Asian monsoon.

In the 1920s and 1930s, Sir Gilbert Walker and E. W. Bliss, while studying the Indian monsoon, further documented this oscillation using a full forty years of station data. Walker named it the Southern Oscillation. He devised an index that later became known as the Southern Oscillation Index (SOI). It utilized a complicated set of equations that used weighted station data for 21 stations not only in the Pacific and Indian Oceans but also in diverse locations such as Africa and South America. He also noted that there was a high correlation between the index in the June through August period to the following December through February period. This led Walker to correctly observe that the index would be useful in long-range forecasting (Walker, 1928).

Around the same time Walker and Bliss were documenting the pressure oscillation, Brooks and Brady, in 1921, were finding another atmospheric relationship associated with El Niño, the relationship between rainfall, winds, and temperatures. They found that the trade winds alternately strengthened and weakened, and with the trade wind changes there were shifts in rainfall location and intensity. Although the wind and rainfall were closely related to the pressure oscillation of Walker's Southern Oscillation, Brooks and Brady did not, at the time, make that connection. According to Mock (1981), it was not until 1933 before J. B. Leighly put the two atmospheric pieces together. He related the changes in rainfall to changes in the trade winds, which in turn he related to the pressure changes of the Southern Oscillation.

Parallel to this work on the atmospheric oscillation, a strong oceanic El Niño warm event in 1925 devastated the Peruvian populations and ecosystems. Peru was then motivated to start gathering rainfall and other weather information to go along with the sea temperature data. But still, researchers looking at the Southern Oscillation paid little notice.

The next big change came in the 1952, when Willet and Bodurtha devised a much simpler formula for the SOI focusing on two locations (Darwin, Australia and the South Pacific island of Tahiti). The new index correlated very well with the more complex Walker SOI. In their work, they reaffirmed the potential value of the SOI in long-range forecasting, again noting the high correlation of the June through August SOI with the SOI during the following winter (December through February).

The atmospheric oscillations were now well documented and widely known. Still surprisingly, no one had yet proposed that the Southern Oscillation and the El Niño phenomenon were interrelated.

LINKING THE SOUTHERN OSCILLATION AND EL NIÑO

The first published attempts to link the atmospheric and oceanic effects occurred in the 1950s and 1960s in papers attempting to find the cause for the Southern Oscillation. In 1956, Schell looked at local sea temperatures and their effect on the South Pacific high pressure center, the strength and position of which determines the strength of the trade winds. He proposed that the "oscillatory" nature of the Southern Oscillation could be the result of a cyclic feedback process.

He speculated that a strengthening of the South Pacific high pressure system would increase the upwelling of cold water off of South America and drive more cold water north then west along the equator, eventually leading to colder temperatures and a weakening of a low pressure in the Indian Ocean. This would reduce the outflow at high levels back toward the east, which would lead to a weakening of the high pressure. This would then lead to a weakening of the trade winds and less cold water upwelling.

However, the whole story was not assembled until the late 1960s when the legendary Jacob Bjerknes (1969), of the University of California at Los Angeles finally associated the oceanic warming of El Niño with the pressure changes, weakened trade winds, and enhanced rainfall of the low-index Southern Oscillation. Dr. Bjerknes had, five decades earlier as a graduate student, gained fame by being the first to clearly define the various stages in the life cycle of mid-latitude storms. Once again, Bjerknes was now piecing together another important phenomenon, the El Niño Southern Oscillation.

Working with data gathered during the International Geophysical Year (IGY), Bjerknes noted that the significant oceanic warming of the strong 1957–1958 El Niño coincided with a strong negative phase of the Southern Oscillation. Thus ENSO was born. Although it was not one of the designed experiments or goals of the IGY, it was a major milestone and proved the value of this kind of organized, wide-ranging observation effort.

Bjerknes proposed the link between the ocean and atmosphere in the form of a direct circulation, which he named the Walker Circulation. He noted in the so-called "cool phase," that the air above the cool water in the eastern Tropical Pacific moves westward toward the warm western Pacific. There the air is heated and moistened over the

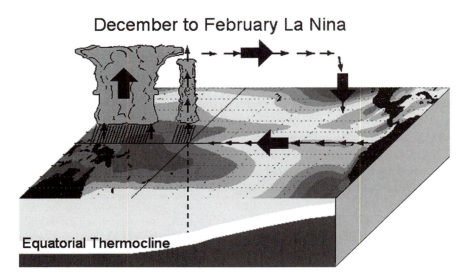

Figure 1.1. "Cool-phase" (La Niña) Walker Circulation. Strong easterlies enhance cool water upswelling off South America and push cold water west along the equator. The combination of cool water and sinking air limits percipitation in eastern areas. Rising air clouds and precipitation are concentrated over western areas.
Source: Climate Prediction Center ENSO Cycle web site http://www.cpc.ncep. noaa.gov/products/analysis_monitoring/ensocycle/enso_schem.html
Courtesy of NOAA

warmer water. The air then rises in showers and thunderstorms and flows back eastward at high elevations (see Figure 1.1).

Bjerknes described the process as a *chain reaction*. He proposed that an increase in the Walker Circulation enhanced the upwelling and the associated cooling of the ocean in the eastern Tropical Pacific, which, in turn, increased the temperature and pressure gradients that caused the flow to begin with.

He also proposed that similarly, a decrease in the low-level easterlies would diminish the upwelling of cold water and decrease the temperature and pressure gradients, causing the Walker Circulation to slow down and even shift. The rising motion may transition to the central or even eastern Tropical Pacific. The air moves westward at high levels of the atmosphere and then sinks over the western Tropical Pacific. In place of the equatorial surface easterlies in the western Tropical Pacific, westerly winds blow to complete the warm phase circulation (see Figure 1.2).

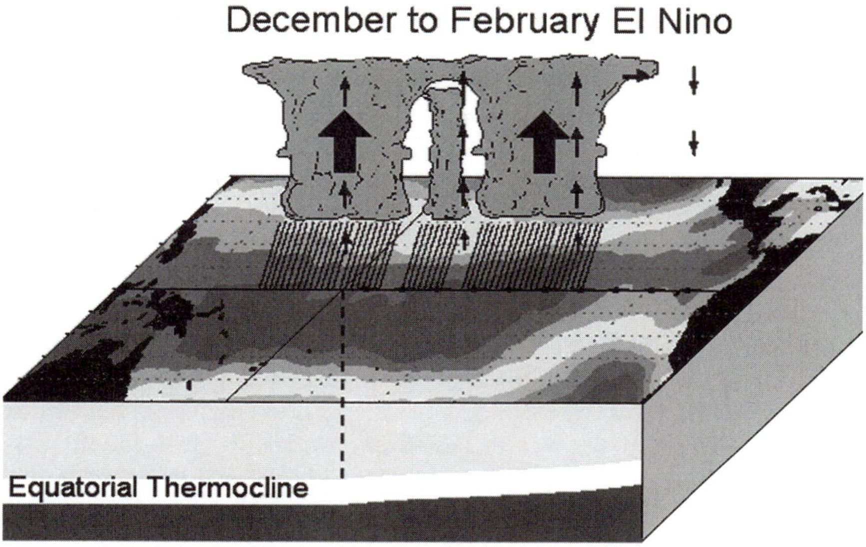

Figure 1.2. "Warm-phase" (El Niño) Walker Circulation. Weakened equatorial easterlies (often becoming westerly winds in western and central areas) reduce cool water upwelling in the ocean off South America. Warm water flows from the west in waves. Rising air clouds and precipitation shift to central (and in strong events to eastern) areas of the Tropical Pacific. Sinking air under high pressure produces drought conditions in western areas.
Source: http://www.cpc.ncep.noaa.gov/products/analysis_monitoring/ensocycle/ meanrain.html
Courtesy of NOAA

One more discovery helped link the oceanic and atmospheric changes in ENSO. In the 1970s, Klaus Wyrtki, an oceanographer at the University of Hawaii discovered the interrelationship between changes in the Pacific Equatorial countercurrent and ENSO. He used a network of tide gauges in the Tropical Pacific, which provided records of sea-level height.

Wyrtiki (1976) noted that El Niños were associated with a transfer of warm water from west to east and a rising of sea levels in the east and a falling in the west. In the cool phase on the other hand, increasing easterly winds lead to a net transfer of water westward with falling sea levels in the east and rising sea levels in the western Tropical Pacific. The movement of water westward increases the upwelling of cold water, which further enhances the Walker Circulation.

The dynamic sloshing of the water in the Pacific basin was accomplished by internal waves called Kelvin waves and Rossby waves. A Kelvin wave is a very long (thousands of kilometers in length) wave that moves east transporting warm western Tropical Pacific water to the eastern Tropical Pacific. When Kelvin waves reach South America they can generate long-lived Rossby waves, which travel west and ultimately cause the demise of an El Niño event.

This dynamic sloshing of water in the Pacific basin is linked with changes in the circulations of air, and thus pressure systems in the atmosphere. It is now known for the first time, truly the degree to which the atmosphere and ocean acted in concert in these events.

The true global implications of the ENSO were still a bit longer coming. Walker had first noted in the 1920s that the low-index (warm El Niño) phase was usually associated with drought in Australia, Indonesia, India, and parts of Africa. He also noted that it also seemed to be related to mild winters in western Canada. Walker was criticized by his colleagues for proposing such far-reaching links. He suggested that the global link and the causes would be forthcoming. He was exactly right, but it took more than half a century for anyone to realize it.

In 1972–1973, a strong El Niño captured the attention of the international press and some researchers because of its impact on the Peruvian fishing industry. But it was the Great El Niño of 1982–1983 that really moved El Niño from the technical journals to the popular media and research institutions. Worldwide catastrophes accompanied the great warming that year. There was major flooding in Peru, Bolivia, Ecuador, Cuba, and the U.S. Gulf States. Rare hurricanes struck Tahiti and Hawaii. Devastating drought and fires scourged sections of southern Africa, southern India, Sri Lanka, the Philippines, Indonesia, Australia, southern Peru, western Bolivia, Mexico, and Central America. The economic losses exceeded $8 billion.

Extensive scientific research followed over the next decade, which confirmed Walker's view that the Southern Oscillation would have global links. Ropelewski and Halpert (1986) of the U.S. National Weather Service's Climate Analysis Center found significant relationships between precipitation and the two phases of the ENSO in many diverse areas of the globe. They discovered that the effects were very nearly the opposite in one phase from the other.

That research has led to many other studies, which have identified the climate effects of ENSO by month, season, and region. With each new event, we learn more about the phenomena, their cause and effects (Rasmusson 1984, 1985; Rasmusson and Wallace 1983; Trenberth 1991; Wallace and Vogel 1994).

FORECASTING EL NIÑO

The first successful numerical simulation and forecast of an El Niño event was made by Cane and Zebiak of Lamont Observatory in 1985. Since that time, numerous climate and statistical models have been developed that forecast ENSO conditions months into the future. Together with oceanic and atmospheric remote-sensing satellites, and a dense network of buoy and ship observations, a system is in place that is capable of observing conditions in real time, and providing 3- to 9-month advance warning of significant shifts in ENSO and in the associated global precipitation and temperature patterns.

"The observations of the climate system, combined with sophisticated ocean-atmosphere prediction models, and the scientific community's increased understanding of the atmospheric response. This led to an incredibly bold forecast of El Niño nearly six months prior to the onset of the major impacts," according to Ants Leetma, Director of the Climate Prediction Center. He continued, "With this event, we were light years ahead of the last major El Niño." Forecasting of ENSO is discussed in greater detail in chapter 8.

Chapter 2

What Causes El Niño and La Niña?

As we discussed in chapter 1, El Niño and La Niña are the two modes of a coupled atmospheric and oceanic oscillation. It is an integral part of the continually evolving global circulation of winds and ocean currents. Changes in the global atmospheric circulation can cause an El Niño or La Niña; and El Niño and La Niña, in turn, can cause changes in the global circulation, at first reinforcing the phenomena and later sowing the seeds of its own demise.

THE GLOBAL CIRCULATION

Global weather systems are driven by global variances in heating and cooling. The sun is the primary driver for these circulation systems and the principal energy source for our planet. The most intense heating occurs in the tropics, because the sun is highest in the sky there and the rays most concentrated when they reach the surface. In higher latitudes, the sun's energy, even during the long summer days, is much smaller and more diffuse when it reaches the surface because the sun is lower in the sky and because the path length of the sun's rays through the atmosphere is greater. The result is that there is much more warming in the lower latitudes than in the higher latitudes. The effect is enhanced in the winter because the days are shorter and the sun's angles even lower. This differential heating induces changes in pressure.

Atmospheric pressure is the weight of a column of air from the surface to the top of the atmosphere. Cold air is denser and thus heavier than a similar volume of warm air. Because of the differential heating with latitude, there is a tendency for pressures at the surface to be higher in the higher latitudes where the air is cold than in the lower latitudes where the air is warm.

The wind blows when there is a pressure variance horizontally. This is because the atmosphere is fluid and seeks an equilibrium state. Air moves from high pressure to low pressure in order to equalize the pressure at the surface. If the earth was uniform and kept the same face to the sun at all times, the surface winds would flow from the poles to the equator. There would be a return flow of air poleward at high levels of the atmosphere, in effect, creating a large circulation.

However, the earth's surface is not uniform, but has alternating land masses and oceans. Land warms and cools much faster than the oceans leading to latitudinal differences seasonally. It should be noted also that the ocean, like the atmosphere, is not at rest. Warm and cold currents of water move with help from the winds. They carry warmer water to high latitudes and cool water to lower latitudes, thus moderating climates downwind.

The fact that the earth rotates on its axis is also very important to weather systems. This results in days and nights. It also causes the idealized hemispheric scale circulation to become unstable and break up into three circulation cells.

The circulation cell nearest the equator is called the Hadley cell. It is the most important feature for ENSO. At the poleward end of the Hadley cell is the subtropical high belt. This used to be called the "horse-latitudes" in the days of the clipper ships. The ships often found themselves becalmed if they came too close to the center of this high-pressure system.

Air sinks in this high pressure, resulting in clear skies and very warm temperatures. The air then flows out of this high pressure, both toward the poles and the equator. The winds that blow from the subtropical high to the equator are called the trade winds. Although the sailing ships of trade wanted to avoid, at all cost, the calm winds near the subtropical high, they did seek out the trade wind belt because the trade winds were very steady and dependable.

The trade winds are deflected toward the west by the earth's rotation. This deflecting force is called the coriolis force and it affects all objects in motion relative to the earth. Both air and ocean water in motion are influenced, deflected to the right in the northern hemisphere and to the left in the southern hemisphere. As a result of both the pressure and coriolis forces, in the northern hemisphere, the trade winds blow from the northeast and in the southern hemisphere the southeast. These trade winds converge on a belt of equatorial low-pressure zone called the Intertropical Convergence Zone (ITCZ). Here the winds blow from the east (called the tropical easterlies) and there are widespread clouds and thunderstorms. This cloudiness and precipitation is supported by the convergence of air from both hemispheres in the trade wind belt. The air is forced to rise and cool and moisture condenses into clouds and falls as rains.

At high levels, the air turns toward the pole where it converges with air moving toward the equator from higher latitudes. This high-level convergence of air "piles up" air in the subtropics and helps maintain the zone of high pressure.

HOW ENSO RELATES TO THE GLOBAL CIRCULATION

The ENSO cycle is linked to changes in the Hadley cell and trade winds. The changes in the atmosphere and ocean are truly "coupled," meaning the atmosphere and ocean influence each other in a very continuous, cyclical way.

In the prelude to La Niña, pressures rise to above normal in the subtropical belts of one or both hemispheres in the east central Pacific. Increased easterly trade winds then act to transport surface water westward where it accumulates in the western Pacific in the vicinity of Indonesia. The sun and air warm the water during its travel so the pool of water that accumulates in the western Pacific typically becomes warmer than normal.

In the eastern Pacific, the waters are normally very cool for the tropical latitude location due to the Peru or Humboldt current. The Peru Current is the southern hemisphere counterpart to the cold California Current of the north Pacific. The cold temperature of the water is enhanced by upwelling of deep-ocean water caused by drag on the ocean by the southeast trade winds and a turning to the left of the water due to the earth's rotation (coriolis force). These factors push water away from the land, thus drawing up cold water from beneath to replace it. During La Niña, stronger trade winds increase the upwelling of cold water along the South American coast and then transport the cold water westward along the equator, where trade winds from both hemispheres converge.

In the oceans, water temperatures generally decrease with increasing depth much in the same way that temperatures decrease with height in the lower atmosphere. In the tropics, the drop of temperature is especially strong as the sun-warmed ocean surface lies above very cold, deep water that flows from the polar icecap regions. The boundary between the warm and cold water is concentrated in a layer called the thermocline. The thermocline depth varies in the oceans due to a number of factors. One of the factors is ENSO.

In the oceans, stronger than normal easterly trade winds literally push water westward. This causes sea levels to become higher than normal and thermoclines deeper than normal in the western Tropical Pacific, and sea levels to become lower than normal and thermoclines shallower than normal in the eastern areas.

The warm water in the western Tropical Pacific warms the lower atmosphere and makes the air unstable. Convection (rising buoyant air currents) is enhanced in the unstable air and as air rises, clouds, heavy showers, and thunderstorms develop. This convection transfers heat upward from the ocean to the atmosphere. As the air warms, the density of the air falls. As the air density falls, pressures drop. This is why pressure in the western Tropical Pacific averages below normal during La Niñas.

Further east, pressures stay higher than normal because both the ocean and air are cooler. This low-pressure in the west and high pressure in the east help maintain stronger easterly trade winds, which further support the La Niña pattern. Air rises in the low-pressure zone in the western Pacific, moves east at high levels, and sinks in the central and eastern Tropical Pacific. The easterly trade winds complete the circulation in the lowest levels. This circulation was first proposed by Bjerknes (1966). He named it the Walker Circulation, after Sir Gilbert Walker who first discovered the Southern Oscillation and its effects.

The enhanced easterly surface flow that is part of the Walker Circulation builds up a reservoir of warm water in the western Pacific. Eventually, this warm pool gets so large that it begins to slosh back eastward into the central equatorial Pacific. The warm water transfer comes in the form of Kelvin waves—subsurface waves of water that travel along the thermocline along the equator.

This change may be enhanced or perhaps triggered by what are now called Westerly Wind Bursts, a short duration low-level westerly wind event that occurs near the equator in the western Pacific and sometimes the Indian Ocean. These reversals of the surface wind from east to west may be brought about by the Madden Julian Oscillation, a periodic oscillation in winds at the surface and at the tropopause (the top of the atmospheric layer where most of our weather occurs). It is especially important as a weather maker in the Indian and western Pacific where it can often be seen operating on a period of between 30 and 50 days, especially in the fall and winter.

Strong westerly bursts may also occur in the western Pacific and Indian Oceans when there is a nearly simultaneous formation of tropical cyclones at nearly the same longitude both north and south of the equator (around 10 degrees latitude). Because the winds blow counterclockwise around cyclones in the northern hemisphere, and clockwise in the southern, a strong westerly wind burst usually occurs in between at the equator.

There are many more westerly wind bursts than El Niños, so it is clear these westerly wind bursts do not always trigger an El Niño. It appears often they simply may help to accelerate one when other conditions favor their formation, or they may help to make the events stronger than they might otherwise become.

Accompanying the eastward expansion of warm ocean waters during El Niño's formative stages, pressures begin to fall and trade winds weaken over the central equatorial Pacific to near the International Dateline. Convection, with its clouds and rainfall, shifts east with the warm water. The Walker Circulation reverses. The air rises in the central Pacific and moves westward only to sink in the region of anomalously high pressure in the western Tropical Pacific. Westerly winds replace the easterly trade winds in at least the western half of the Tropical Pacific.

Further east, even if the trade winds continue to blow from the east, they weaken. The upwelling of cold water along the South American coast decreases and the sea surface temperatures rise. The warm water sloshing eastward from the western Tropical Pacific continues advancing eastward and eventually (usually a few months later) reaches eastern areas and help further enhance the warming and deepen the warm water layer. The El Niño now is well underway. In stronger El Niños (such as 1982–1983 and 1997–1998), the convection and rainfall can shift all the way to the South American coast.

The warm water "plume" at this stage is clearly seen along the equator from the South American coast to the vicinity of the International Dateline. Heat is transferred to the atmosphere through the convection process. It may take a year or more to deplete the excess heat through this process. As the ocean temperatures cool, pressures rise in the central Pacific and the trade winds again increase. Sea levels rise in the west and begin to fall in the east. Upwelling brings cold subsurface water up to the surface in the eastern Tropical Pacific where the equatorial easterlies push it west along the equator. The La Niña has begun (see Plate 1). Peixoto and Oort (1992) found that the full development of the sea surface temperature warm and cool pools may lag up to four and one-half months from the first pressure and SOI changes.

These large-scale coupled shifts in warm water pools and atmospheric pressure, rising and sinking air and associated cloudiness and rainfall, act to redistribute the stored ocean heat. The excess heat that is transferred to the atmosphere during El Niño gets carried by atmospheric winds and storms to high levels of the atmosphere and all parts of the globe, and eventually back out to space.

The heating has an effect on weather patterns in mid and high latitudes, favoring particular configurations of the jet streams. Because the cool and warm water pools and associated clouds and rainfall patterns are in opposite locations in El Niño and La Niña, it should not be surprising that the global climate anomalies associated with El Niño tend to be the opposite of those with La Niña.

OTHER POSSIBLE EL NIÑO TRIGGERS

In addition to the theories about the importance of the periodic sloshing of water in the Pacific basin and the Kelvin and Rossby waves, there are other theories about other conditions, which may trigger the onset of El Niño. Rothstein (1996) summarized some of them.

White (1964) related the build-up of the tropical ocean heat content to the onset of El Niño.

Hirano (1988) and then Nicholls (1990), Kirchner (1995), Robock (1995) and Portman (1996) studied the theory that low-latitude volcanic eruptions may trigger the onset, presumably by altering the Hadley cell circulations (and thus the strength of the equatorial easterlies).

Fairbridge (1990) saw signals in the solar and lunar cycles, whereas Walker (1995) found a possible connection to siesmicity in the ocean floor of the Pacific.

Finally, NASA Jet Propulsion Lab scientists and researchers at the University of Washington (Mantua 1997) proposed that a longer term, larger-scale oscillation in the Pacific (PDO or Pacific Decadal Oscillation) predisposes the ocean and atmosphere towards either El Niño or La Niña.

MID-LATITUDE EFFECTS

In the tropics and subtropics, the main weather players are the sub-tropical highs and the trade winds. The highs and trade winds weaken (El Niño) and strengthen (La Niña) in a see-saw phenomenon known as the Southern Oscillation.

In mid-latitudes, the main weather players are the westerlies, the jet streams and the polar front. The jet stream is a high-speed current of air at upper levels of the atmosphere (usually 6 to 8 miles high). The polar jet stream reflects the region of the strongest contrast in temperatures in lower levels. This usually coincides with the polar front, which is the boundary at the surface between the cold air masses of polar origin and warmer, sometimes tropical air. There is also a subtropical jet stream found in the subtropics representing the boundary between true tropical air and much modified air from the polar regions. Often in winter, both the polar and subtropical jet streams can be found on the weather maps.

The polar jet stream undulates alternately north and south in Rossby waves. The location and amplitude of the waves determines what type of weather is experienced as they influence where storms develop and move. As indicated, earlier, the polar jet stream coincides with the polar front at the surface. Storms move along the polar front gaining a great deal of their energy from the jet stream. The storms also derive some energy from heat and moisture from the tropics. This effect is maximized when the subtropical jet stream is also in-volved in a strong storm.

The El Niño and La Niña each favor a different location for the dips and bulges of the polar jet stream. They also affect the strength of the subtropical jet stream. In this way, they influence the weather in middle latitudes. The influence is greatest in the winter months when the coupling of the tropical and mid-latitude patterns is the best (the jet stream and storms associated with it are closest to the warm and cold water pools of ENSO).

It should be noted that no two ENSO events are perfectly alike.

The sometimes subtle differences in the strength and position of the warm and cold pools and other factors can alter the pattern from the mean, sometimes significantly. The western region of the National Weather Service repeatedly communicates to the media that the impact of the ENSO phenomenon is more about probabilities than predictions of specific storms.

THE EL NIÑO PATTERN

In most El Niño winters, the warming of the air due to strong convection over warm water over the eastern and central Tropical Pacific helps energize the polar and subtropical jet streams to the north. A strong low pressure develops over the Aleutians. The polar jet stream curves to the north into northwestern North America, while the subtropical jet stream ripples across northern Mexico or the southern United States. Strong Pacific storms that normally come ashore in the Pacific Northwest, instead move into Alaska. Fewer than normal storms take the northern route across the northern United States or southern Canada; and those that do tend to be weaker than normal.

On the other hand, more storms than normal ride the supercharged subtropical jet stream into California and then redevelop along the Gulf and/or east coasts of the United States. These storms can bring heavy, flooding rains. They become especially strong when disturbances riding the polar jet stream also become involved (when the polar and subtropical jet streams merge).

The strong and active southern storm track means heavy precipitation there, while the absence of the northern storm track means winter dryness (and thus less snow than normal) in normally wintry northern locations.

THE LA NIÑA PATTERN

During La Niña winters, on the other hand, the polar jet stream is strong and the subtropical jet stream weaker in the vicinity of North America. Stronger than normal high pressure in the central Pacific deflects the jet stream further north, thus concentrating temperature contrast and accelerating the jet stream winds. Meanwhile to the south, the subtropical jet stream is weaker because there is less warming over the cooler than normal water in the tropics and the contrast in temperatures from the tropics to the subtropics is less.

Sometimes in La Niña winters it may be hard to even find the subtropical jet stream. The polar jet stream typically dips into the western and central United States and then turns north near the eastern states. Storms come in rapid-fire succession into the Pacific Northwest, battering the coast with strong winds and heavy rains and bringing extremely heavy snows to the mountains. These storms then travel across the northern states and southern Canada bringing moderate to heavy snowfall. The absence of storms on the southern route across the nation often means a winter drought for the south.

Chapter 3

The Global Effects of El Niño and La Niña

In the 1920s, Walker proposed that the negative phase of the Southern Oscillation known as El Niño might have an effect on weather in such remote locations as western Canada and predicted that some day its global nature would be understood. Although this was not taken seriously at the time, today it is clear that Walker was indeed correct.

Both the warm El Niño and the cold La Niña have been shown statistically to have global climate implications. The effects are very nearly the mirror opposite of one another. In both cases, large-scale shifts in the position of the jet streams and major weather features results in areas that are excessively wet or excessively dry.

El Niños produce wetter than normal conditions across southern Brazil and Argentina, the southern United States, east-central Africa, southernmost India and Sri Lanka, the central Tropical Pacific Islands, and in strong events parts of Peru and Ecuador (see Figure 3.1).

El Niño produces widespread dryness and drought across Australia, India, Indonesia, the Philippines, Brazil, and Venezuela, parts of east and south Africa, the western Pacific Basin Islands, central America, and western Canada and the northwest and north central United States.

Heavy El Niño rains can fall in desert regions bringing an amazing, although temporary, transformation. In Peru, the Great El Niño of 1982–1983 brought more than 100 inches of rain to arid regions in the northern part of the country, which normally receive less than 6

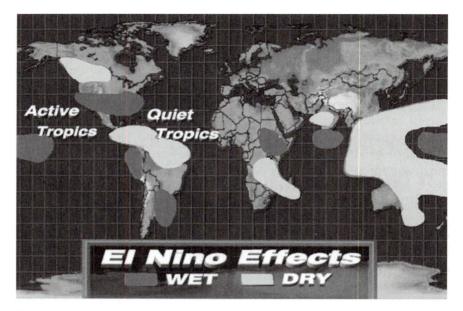

Figure 3.1.   The global effects of El Niño on precipitation. Drought is a problem in many tropical areas that depend on the arrival of the summer monsoon rains, while other normally dry areas receive flooding rains, which can transform deserts into lush grasslands.
Source: Intellicast—http://www.intellicast.com/DrDewpoint/Library/EL/
Courtesy of Intellicast.com

inches of rain each year. These rains transformed a desert into a grassland peppered with lakes. Grasshoppers were attracted to the lush vegetation, which in turn led to sudden increases in the populations of birds and toads. There were some temporary benefits from the changes to the environment. The lush lakes also were home for fish that had been trapped during floods. Shrimp too abounded in the flooded coastal water regions.

In Peru, rains in places exceeded that of the 1982–1983 and the results were very similar. Rains some days falling at a rate of 5 or 6 inches per day, produced massive mudslides and floods. Runoff from the flood poured into the coastal Sechura desert. It transformed the desert wasteland into the second largest lake in Peru (90 miles wide, 20 miles long and 10 feet deep).

But with the good came the bad. Floods in tropical areas can bring an increase in tropical diseases. There was an increase in mosquito-borne malaria in Peru in 1982–1983. In eastern Africa, the 1997–

1998 El Niño floods brought an epidemic of "Rift Valley Fever" with more than 500 deaths and diseases decimated the livestock population.

Heavy El Niño storms can bring floods, mudslides, and even tornadoes. California and the Gulf States, especially Florida are often hardest hit during strong El Niños. Billions of dollars in damage occurred in the Great El Niño of 1982–1983 in these two regions. Record rainfall occurred in parts of California and Florida during the El Niño of 1997–1998. In California, strong storms brought hurricane-force winds, thirty-foot waves, and flooding rains. Thirty-five California counties were declared federal disaster areas.

Fifty-four of Florida's sixty-seven counties were declared disaster areas from flooding rains, tornadoes, and damaging thunderstorm winds. Winds from one severe thunderstorm gusted to 104 mph in Miami. The damage was the worst in Florida since the "Storm of the Century" in 1993, also an El Niño year.

El Niño often enhances tropical activity in the eastern Pacific. These storms threaten the western coast of Mexico and even the southwestern United States. Some storms that affected these areas are discussed in chapter 4.

During an El Niño, monsoon rains fall over the central Pacific instead of the west. Typhoons and hurricanes feeding off unusually warm water are steered off their normal tracks to islands like Tahiti and Hawaii, normally unaffected by these storms. A rare hurricane made landfall in Hawaii in November 1982. In the Great El Niño of 1982–1983, Tahiti and the Islands of French Polynesia were hit by six hurricanes and tropical storms, which did millions of dollars in damage. Hurricane Iniki hit the Hawaiian Island of Kauai in September 1991. It killed seven and left an estimated $1.8 billion in damages. The El Niño year of 1997 brought eight tropical cyclones to the central Pacific.

At the end of the last El Niño in 1998, China suffered through the worst flooding in nearly half a century. The floods were widespread extending to twenty-eight provinces and affecting 230 million people—more than 20 percent of China's population. The floods first were felt across the central and southern parts of the country along the Yangtze and other rivers. Later in the summer, the heavy rains and floods spread to the northeastern parts of the country. 3,656 people died and more than fifteen million were made homeless.

Some 26,440 hectares of farmland were flooded and fifteen million farmers lost their crops. The storm caused serious damage to critical facilities such as health clinics, schools, water supply, and infrastructures like roadways, bridges, and irrigation systems. The total economic losses were estimated at $32 billion (U.S. dollars).

The El Niño droughts can produce a whole other set of problems. Researchers have found the strongest connections between El Nino and intense drought in Australia, India, Indonesia, the Philippines, Brazil, parts of east and south Africa, the western Pacific Basin Islands (including Hawaii), Central America, and various parts of the United States. Drought occurs in each of the these regions at different times (seasons) during an event and in varying degrees of magnitude. Ropelewski and Halpert (1987) also looked at the link between El Niño events and regional precipitation patterns around the globe. Northeastern South America from Brazil up to Venezuela shows one of the strongest relationships. In seventeen El Niño events, this region had sixteen dry episodes. It is not uncommon to find the rain forests burning during these dry periods.

Other areas, from their study, also showed a strong tendency to be dry during El Niño events. In the Pacific Basin, Indonesia, Fiji, Micronesia, and Hawaii are usually prone to drought during an event. Virtually all of Australia is subjected to abnormally dry conditions during El Niño events, but the eastern half has been especially prone to extreme drought. This is usually followed by widespread bush fires and a decimation of crops. India has also been subjected to drought through a suppression of the summer monsoon season that seems to coincide with El Niño events in many cases. Eastern and southern Africa also showed a strong correlation between ENSO events and a lack of rainfall that brings on drought in the Horn region and areas south of there. One final region they found to be abnormally dry during warm events was that of Central America and the Caribbean Islands.

In addition to crop failures, the dry heat in normally wet climates can lead to vast forest fires. This is actually more of a problem in wet climates than dry climates because of the great biomass available to burn. In arid, tropical climates, there is little vegetation and most of it is adapted to dry conditions (drought-resistant plants called Xerophytic plants). In normally wet tropical areas, there is a tremendous growth of trees and shrubs during wet years that can dry up and burn during drought years.

In 1997–1998, the worst drought in Malaysia in 50 years spread fires that claimed more than 1 million acres. Smoke from the fires was a hazard to health and navigation and was responsible for an airline crash that claimed hundreds of lives. Thousands of square miles of forestland also burned in Guatemala, Honduras, Nicaragua, and Mexico. Tens of thousands of miles of forest burned in Brazil. Fires also burned in parts of Africa including Tanzania, Kenya, Rwanda, the Congo, and Senegal.

LA NIÑA

La Niñas produce nearly the mirror opposite of the precipitation effects of El Niño. Dryness occurs across the southern United States, southern Brazil and Argentina, southern India and Sri Lanka, east central Africa and parts of China and the Middle East (see Figure 3.2).

Meanwhile, it is unusually wet across parts of Australia, Indonesia, Southeast Asia, the interior of India, southeastern Africa and Madagascar, from central Brazil to Venezuela and Colombia and much of Central America and the Caribbean.

In the United States, Phillips found a tendency for heat and drought during the summer of La Niña years in the Corn Belt, with an average 5 percent reduction in corn yields. During the summers of El Niño years, there was a tendency for moist and relatively cool conditions with an average 4 percent increase in corn yields (Phillips, Rajagopalan, Cane, & Rosenzweg, 1999).

Similar summer droughts are found in parts of the Middle East, Africa, and China. In China, as a result of the La Niña in 1999 and 2000, a failure of the summer rainy season and high temperatures dried up the Songhua River, threatening the economy of China's Northeast. It ruined thirty-five million acres of crops and left 16.2 million Chinese residents without water. The Songhua River is a source of irrigation for the country's most productive grain-growing areas. These areas were part of the large region that suffered through flooding at the end of the last El Niño in 1998.

During the same two-year La Niña in Iran, the government reported that the drought had seriously affected eighteen of the country's twenty-eight provinces, causing four major lakes to dry up and water shortages affecting twelve million people (60 percent of the population in those regions). Camels were reportedly found dead along roadsides. The drought was the worst in thirty years in Afghanistan. Parts of Pakistan were said to be suffering through their worst drought in the country's history. Although droughts in La Niña

Figure 3.2. The global effects on precipitation of La Niña. The pattern is very nearly opposite to that of El Niño. Heavy monsoon rains occur in the regions that depend on this seasonal rainfall. Drought can occur in other regions, including agricultural regions of the United States and South America.
Source: Intellicast—http://www.intellicast.com/DrDewpoint/Library/EL/
Courtesy of Intellicast.com

years can cause regional problems (see chapter 4 for the drought impact in the United States), for most of the populated tropical areas, if there is a problem, it is too much rainfall.

La Niña years are usually years with reliable monsoon rains. This is mostly good news because it means good crop yields and abundant livestock. However, there is danger from having both too much monsoon rains, which can mean flooding and more frequent and stronger tropical storms, which can do major damage.

La Niña events are usually associated with increased tropical storm activity in the Atlantic. These storms threaten landfall across the Caribbean, Central America, the East Coast of Mexico and the East and Gulf Coasts of the United States (Caviedes 1991; Gray 1984).

Very late in the 1998 hurricane season, on October 26 and 27, meandering Hurricane Mitch became the fourth strongest Atlantic

Basin Hurricane ever with winds exceeding 180 mph. Hurricane Mitch was the strongest storm in the western Caribbean since Gilbert in 1988 (also a La Niña year).

Mitch stalled off the coast of Honduras from late on October 27, 1998 to the evening of October 29 before moving onshore. The storm dumped incredible rains on the mountains of Central America, causing floods and mudslides that were responsible for the death of between 10,000 and 12,000, making it one of the top five deadliest hurricanes on record in the Atlantic Basin.

Hurricane Mitch was another on the list of La Niña Atlantic/Caribbean powerhouses to affect the region. Some other very powerful La Niña-year storms affecting this region included Janet in 1955, Edith in 1971, and the memorable Gilbert in 1988. Other strong systems, which caused deaths and significant damage in the region included Ella in 1970, Carmen in 1973, Joan in 1988, and Roxanne, another meandering storm, in 1995.

Chapter 5 lists some of the many major La Niña storms to affect the east coast of the United States.

In addition to Central America, Mexico, and the United States, La Niña seems to hit areas in Asia hard with tropical storms and flooding. La Niñas, its appears, especially threaten major tropical flooding in the Bay of Bengal in places like Bangladesh and eastern India. Bangladesh is one location that seems especially prone to problems. This lowland country with 80 percent of the land in the floodplain, is subject to flooding with both monsoon rains and rains from tropical cyclones. La Niña increases the summer monsoon rains from May to September in this area and usually results in flooding. La Niña increases the chances of major cyclones later from September to November, which can produce even greater disasters.

The list of years with major flooding in Bangladesh is virtually the list of La Niña years: 1954, 1955, 1956, 1962, 1968, 1970, 1971, 1974, 1984, 1987, 1988, 1998, 1999, and 2000 (all but two of these years were La Niña years.) In some La Niña years, major cyclones (what hurricanes are called in Bangladesh) have inundated the country and produced massive loss of life.

During La Niña events in the early 1970s, one storm produced an incredible loss of 300,000 lives and another storm 30,000 lives. In 1988, two thirds of Bangladesh was left submerged and 2,000 people died. Twenty-five million people overall were affected by the flooding and many were left ill from water-borne diseases.

On October 29, 1999, a massive storm from the Bay of Bengal

bypassed Bangladesh and tore into eastern India's coastal state of Orissa with winds that were sustained at 155 mph and that gusted to 190 mph. The storm surge of more than 20 feet carried a wall of water 13 miles inland. The death toll was estimated by the Red Cross to be more than 10,000. An estimated 300,000 cattle also perished. Many thousands of both people and animals suffered in the weeks afterward from diseases and other storm-related problems. In one tragic example, industrial chemical acids, which leaked into bodies of water, scarred or poisoned many hundreds of survivors who bathed in or drank from the water.

Chapter 4

El Niño and La Niña Effects on Climate in the United States: Winter and Spring

EL NIÑO WINTER WEATHER PATTERNS

In North America, particularly the United States, the impacts of El Niño are most dramatic in the winter. El Niño produces winters that are generally mild in the northeast and central United States and wet over the south from Florida to Texas and often as far west as California. Alaska and the northwestern regions of Canada and the United States can be abnormally warm. This can be attributed to the forcing resulting from a Pacific–North American (PNA) pattern that is typified by a high-pressure ridge over northwestern North America and a low-pressure trough in the southeastern United States. This serves as an upper level steering mechanism for moisture and temperature at the surface. Once the pattern is entrenched, regions under the ridge stay dry while those downstream of the trough get heavy rains or snows and possible flooding.

Ropelewski and Halpert (1986) studied North American precipitation and temperature patterns associated with both El Niño and La Niña conditions and concluded the following. In the Great Basin area of the western United States, above-normal precipitation was recorded during El Niño years in 81 percent of the cases for the "season" that runs from April to October. In the southeastern United States and northern Mexico, above-normal precipitation was also recorded for 81 percent of the cases for the "season" that began in October of the year and concluded in March of the following year.

For temperature anomalies during El Niño conditions in North America, the Pacific Northwest in the United States, western Canada, and parts of Alaska showed warmer temperatures in 81 percent of the years, while the southeastern United States showed below-normal temperatures around 80 percent of the time. This would seem consistent with a typical PNA atmospheric pattern. In the coastal west, the displacement of the jet stream can bring abnormally large amounts of rain and flooding to California, Oregon, and Washington.

During El Niños, the combination of the Southern Oscillation-induced pressure patterns and the warm pool of water in the eastern Tropical Pacific cause a strong vortex of low pressure to develop in the region of the Aleutians during the winter months. Very strong jet stream winds blow south of this vortex. The jet stream is a high-speed ribbon of air that generally blows from west to east in middle latitudes. In winter, the winds at the jet stream level, from 6 to 8 miles up, can blow at 200 mph, equaling the strongest winds ever recorded at the surface during hurricanes! These winds are strongest in the zone of greatest temperature contrasts. That is also where the storms tend to track.

In El Niño winters, the jet stream splits over western North America. The northern branch, called the polar jet stream, lifts northeast into western Canada, while an energized southern branch, called the sub-tropical jet, races across the southernmost United States or even northern Mexico (see Figure 4.1). This flow usually results in milder and drier than normal weather across the western and central provinces of Canada and the northwest and north central United States. Snowfall tends to be below normal in a large area across the northern United States from Washington State to Michigan and the Ohio Valley.

But there are differences between strong (very warm) and weak El Niños. When temperatures in the warm plume are just slightly above normal, we get a similar general jet stream pattern, but there tends to be more cold air masses and thus more snowfall. In strong El Niños with ocean temperatures well above normal ($>1.5°C$) in the eastern Pacific, strong convection may shift all the way to the eastern Tropical Pacific. The surface pressure patterns that accompany this warming also result in less upwelling off the western parts of North America. It is this upwelling of cold water that maintains the normally cool ocean in eastern areas. As a result, typically in strong El Niños, warmer than normal ocean temperatures extend in general well outside the warm El Niño plume along the equator throughout the entire eastern Pacific basin. This results in a warming of the atmosphere

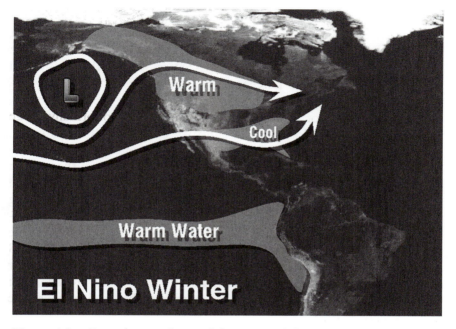

Figure 4.1. Jet stream and typical features and flow during the mild interludes in an El Niño winter.
Intellicast—http://www.intellicast.com/DrDewpoint/Library/EL/
Courtesy of Intellicast.com

downwind over North America. In addition, the strong convection excites an atmospheric pattern that results in strong sinking of air in south central Canada. This produces further warming of the air and a virtual exclusion of arctic air (see plate 2).

Most of the winter storm activity is associated with the southern jet stream, called the subtropical jet stream. This southern storm track is energized by the El Niño and produces strong storms with copious amounts of rain in the southwest. In California, the El Niño threatens heavy flooding rains and mudslides (see Figure 4.2).

In the nine classic El Niños since 1950, southern California (Los Angeles) had above-normal rainfall in all nine years while northern California (San Francisco) had above-normal rainfall in seven years. In the most recent El Niño (1997–1998), Los Angeles had a total of 31.01 inches (210 percent of normal) and San Francisco 47.19 inches (230 percent of normal). Santa Barbara, California set an incredible monthly total in February 1998 of 21.74 inches, the most for any month since 1867.

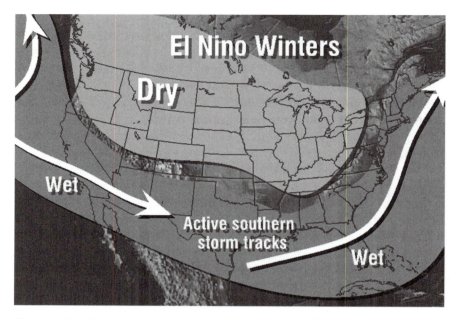

Figure 4.2.   Storm tracks and precipitation during El Niño winters
Source: Author.

Six of the ten costliest California winter storm seasons were El Niño years. The El Niño winter of 1994–1995 produced $1.95 billion damage, and the storms in 1982–1983 produced $1.44 billion. The winter of 1968–1969 brought storms that resulted in $1.34 billion damage. The damages in the winter of 1992–1993 totaled $0.61 billion, and the winter of 1997–1998 $0.55 billion. The southern Rocky Mountains from southernmost Colorado to New Mexico often receive heavy snows during El Niño winters. This extends eastward into west central Texas. Abilene, Texas averages nearly twice as much snowfall during El Niño winters as in non-El Niño winters; and six of Abilene's top nine snowy winters were during El Niño years (McCullough, NWS, San Angelo, TX report on the NWS Web site in 1997).

These storms often redevelop in or near the Gulf of Mexico, where they pick up additional moisture and produce very wet and stormy weather east to the Gulf states and especially the Florida peninsula. The storms often then take the turn up along the east coast to bring heavy rains or snows, strong winds, and coastal flooding.

Florida typically has well above-normal rainfall and bouts of severe weather during El Niño winters. Winter season is normally the driest

and sunniest season in the "sunshine state." During the strong El
Niño of 1997–1998, in December alone, Tampa received 15.57
inches of rain, while Orlando had 12.63 inches.

During the late evening of February 22 and the early morning of
February 23, 1998, a large outbreak of tornadoes occurred across
Florida, causing forty-two fatalities. A total of 800 homes were destroyed, another 700 were left uninhabitable, and another 3,500
damaged. Some 135,000 utility customers lost power. The damage
from the outbreak exceeded $60 million, bringing the total for the
winter to $500 million. Of Florida's 67 counties, 54 were declared
federal disaster areas during the 1997–1998 El Niño winter.

In weaker El Niños, where the warming of the global atmosphere
is more limited, the southern jet stream can result in prolific snowfalls
near the east coast. Kocin and Uccellini (1998) studied major snowstorms affecting major metropolitan areas in the eastern United States
from 1950 to 2000. They identified 28 major storms during that period. Sixteen of the storms occurred during El Niño winters, nine in
neutral winters, but only three during La Niña winters (see Table 4.1).

In addition, some of the "legendary" heavy snowfall winters of the
20th century in the east have been in El Niño winters (1957–1958,
1968–1969, 1976–1977, 1977–1978, 1986–1987, 1987–1988, and
1992–1993). On the other hand, some of the strongest (warmest) El
Niños produced so much large-scale warming that the inevitable train
of storms brought only rain to the coastal cities. Examples of the
stormy but snowless El Niño winters included 1972–1973, 1991–
1992, and 1997–1998. Ironically, the Super El Niño of 1982–1983,
produced one major blizzard (called the Great Megalopolis Blizzard).
This storm buried major cities from Washington to Boston under 2
feet of snow. Some climatologists have suggested that the volcanic
eruption of El Chichon earlier that year may have produced sufficient
atmospheric cooling to offset the El Niño warming just enough to
allow precipitation during this event to fall as snow. It may be that
the storm just managed to time its track up the east coast to one of
the few cold air masses of the season.

NOAA's Midwest Climate Center found that in the Midwest, El
Niños bring a reduction of snowfall on the order of 10 to 20 inches
for many areas. The areas with the largest reduction included northern Illinois and Indiana, in Michigan along the shores of Lake Michigan, and in the Upper Peninsula, west central Minnesota, southeast
North Dakota and northeast South Dakota, and the eastern shores
of Lakes Erie and Ontario.

Table 4.1
Major East Coast Snowstorms and
ENSO State

Blizzard	ENSO
March 18–19, 1956	La Niña
February 14–17, 1958	El Niño
March 18–21, 1958	El Niño
March 2–5, 1960	Neutral
December 11–13, 1960	Neutral
January 18–21,1961	Neutral
February 2–5, 1961	Neutral
January 11–14, 1964	El Niño
January 29–31, 1966	El Niño
December 23–25, 1966	Neutral
February 5–7, 1967	La Niña
February 8–10, 1969	El Niño
February 25–28, 1969	El Niño
December 25–28, 1969	El Niño
February 18–20, 1972	Neutral
January 19–21, 1978	El Niño
February 5–7, 1978	El Niño
February 17–19, 1979	Neutral
April 6–7, 1982	Neutral
February 10–12, 1983	El Niño
January 21–23, 1987	El Niño
January 25–27, 1987	El Niño
February 22–24, 1987	El Niño
March 12–14, 1993	El Niño
February 8–12, 1994	Neutral
February 2–4, 1995	El Niño
January 6–8, 1996	La Niña
March 31–April 1, 1997	La Niña

Source: P. J. Kocin and L. W. Uccellini, Snowstorms along the northeastern coast of the
United States, *Meteorological Monographs*
44, 1990, 280 pp. Courtesy of Meteorological Monographs.

LA NIÑA WINTER WEATHER PATTERNS

La Niñas, have nearly the mirror-opposite effects on global weather
patterns to El Niño. This is especially true in winter. Whereas most
El Niño storms take the southern track across the United States, in
La Niña years, the storm tracks are mostly across the northern states

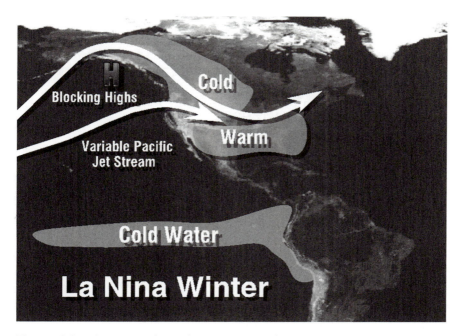

Figure 4.3.   Storm tracks and precipitation during La Niña winters
Source: Author.

(see Figure 4.3). Not surprisingly, that is where most of the snows fall in La Niña winters. The Pacific Northwest is especially vulnerable. In general, this region receives nearly double the amount of snow during La Niña winters than in El Niño winters.

The heavier than normal precipitation extends down to northern California. Southern California usually receives less than normal rainfall during La Niña winters. Eight of the nine La Niñas since 1950 had below-normal rainfalls in southern California. In San Francisco, the average La Niña rainfall is very near normal (see Figure 4.4).

Most of the snowfall records in this region were set in La Niña winters. The Mt. Baker ski area in northwestern Washington State reported 1,140 inches of snowfall for the 1998–1999 snowfall season ending June 30, 1999. This was a new world record for seasonal snowfall. The previous world seasonal snowfall record was 1,122 inches during the 1971–1972 La Niña at the Paradise Ranger Station on Mt. Rainer, also in Washington State and about 150 miles south of Mt. Baker. Stevens Pass and Snoqualamie Pass set their all-time records in the La Niña winter of 1955–1956.

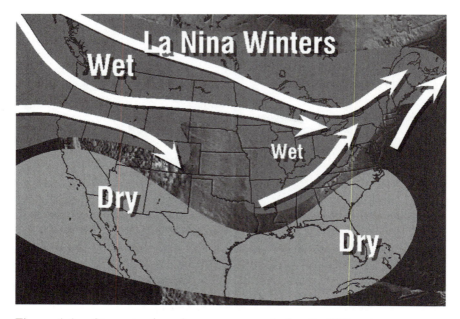

Figure 4.4.   Storm track and temperatures during La Niña winters
Source: Author.

Snowfall tends to be above normal north of a line from northern California through central Utah and Colorado. It tends to be below normal south of that line. Snowfall tends to be above normal in the Northern and Central Plains states across the Great Lakes and Ohio Valley (and in some years even the Tennessee Valley) and northern New York and New England. The snows over the Great Lakes come from a combination of storms that converge on the Lakes and from lake effect snows from cold air that feeds across the Great Lakes during the cold phases of La Niña winters. During the weak La Niña of 1996–1997, the Northern Plains received record snowfall, which led to record spring flooding when snowmelt occurred.

La Niña winters often produce significant variability. There are often lengthy spells of real winter with frigid arctic air, subzero cold, and widespread snow and ice. These spells can last 2 to 4 weeks in a typical La Niña year. They are usually followed by a turn to unusually mild conditions, with Pacific air overspreading the country bringing thawing temperatures, the risk of heavy rains, and if there is important snowmelt, possible flooding.

It all relates to the configuration of the jet stream. The jet stream

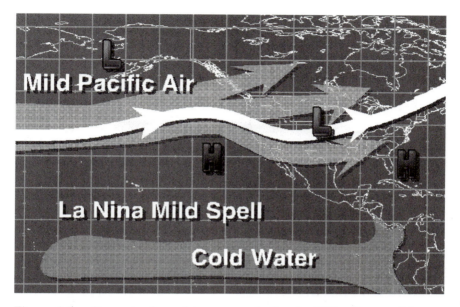

Figure 4.5.   Storm tracks and jet stream during La Niña winter mild spells.
Source: Author.

is not a steady wind. It changes position, direction, and speed continuously as waves move through the atmosphere (very much like waves in the ocean do). The changes in La Niña winter make a huge difference in the kind of weather we experience here in the United States.

There are times when the high-level jet stream winds blow primarily west to east (Figure 4.5), which meteorologists refer to as a zonal flow. When this flow occurs here in the in United States, Pacific low-pressure systems, often heavily laden with moisture, ride the jet stream ashore in the west bringing heavy rains to the coast and heavy snows to the mountains. The Pacific storms also transport with them relatively mild Pacific air. This air gets further warmed from heat released from the condensation process during ascent over the western and Rocky Mountains, and then from compression as it sinks down to the Plains States. This pattern brings the thaws during La Niña winters.

When this flow is flat across both the lower 48 states and Canada, the cold air supply can be shut off entirely. The ripples in the flow from the Pacific systems then bring rains to even some northern lo-

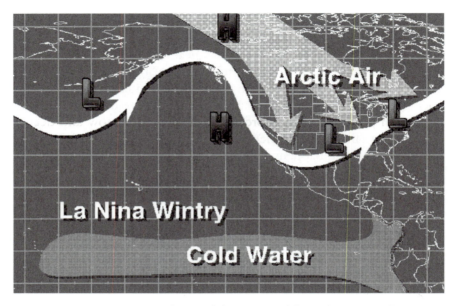

Figure 4.6.   Jet stream and typical features and flow during the first
stages of the wintry periods in a La Niña winter.
Source: Intellicast—http://www.intellicast.com/DrDewpoint/Library/EL/
Courtesy of Intellicast.com

cations. If the rains are heavy enough, and there is a lot of snowmelt,
heavy flooding can result.

These spells are generally good news for snow lovers in the West,
where the mountains are high and the air is cold enough for snow.
But this pattern is bad news elsewhere. Fortunately, the warm spells
are usually then followed by equally long periods of good or even
great winter sports conditions. This is because the jet stream does not
stay flat for very long. When it begins to buckle, it bends well to the
north and then turns and bends equally far to the south. In La Niña
years, this buckling tends to occur in certain prescribed locations (Figure 4.6).

The northward bend in the jet stream tends to occur in the easternmost Pacific, while the downstream dip typically sets up in the
western and central parts of North America. The jet stream then turns
northeast usually across the Mid-Atlantic States and out into the Atlantic.

When this buckling happens in this location, arctic air masses begin to quickly build in northwestern Canada. This frigid air is steered to the south and east by the jet stream winds. Storms ride the leading edge of the cold air and bring rains ahead and snows behind. The cold air can come down west of the Rockies at times and bring frigid conditions to the Intermountain region and sometimes even to the Pacific Coast. The cold air usually comes down from the source regions in pieces. Each reinforcement brings more snow and deeper cold. Eventually the "mother lode" of cold may descend into the states. When that happens, subzero cold is widespread, sometimes even coast to coast.

Even more significant snows may occur in areas bordering this last mountainous high pressure. Along the very edges of the very cold and very dense air mass, the cold air is typically shallow. Warmer air aloft often makes ice storms a threat in this region, which can be a real hazard for travelers. Ironically, it is often worse in the lower elevations. Aloft in the warmer air, the slopes and mountaintops may see only rain or drizzle.

Winter Wild Card: The North Atlantic Oscillation

Despite the high reliability of La Niña and El Niño patterns in winter, there are always wild-card factors that could affect the patterns in significant ways. One such wild-card factor is the North Atlantic Oscillation (NAO). Pressures in the North Atlantic tend to vary north to south in one of two ways. When pressures are below normal to the north (in the vicinity of Iceland or Greenland), pressures to the south near Portugal or the Azores tend to be unusually high. This is called a positive or high index situation. When pressures, on the other hand, are unusually high to the north, they tend to be unusually low to the south. This is a negative or low index (sometimes called a blocking) situation.

When the NAO is positive, a strong westerly flow drains cold air off the North American continent and above-normal temperatures and below- to much below-normal snowfall often result in most of the eastern half of the nation. When the NAO turns strongly negative and blocking patterns develop, cold air gets bottled up over North America and the storm tracks are deflected south in the eastern states. This typically moves the heavy snow belts south.

The La Niña winter of 1988–1989 was a La Niña winter with a strong positive NAO. It was a dreadful winter for snow lovers. On

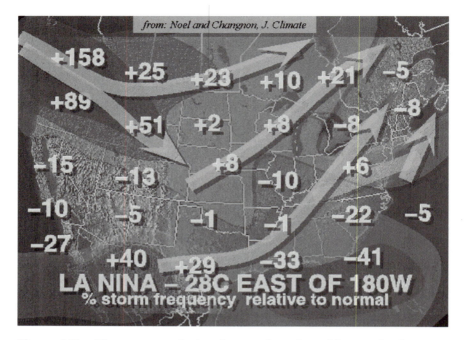

Figure 4.7. The storm tracks in winters when the cold water in the eastern tropical Pacific was confined to easternmost areas of the tropical Pacific. Note the high frequency of storms in the Pacific Northwest.
Courtesy of Intellicast.com from data by Noel and Changnon (1998)

the other hand, the incredible winter of 1995–1996 was a weak La Niña winter with the NAO strongly negative. Many of the major metropolitan areas in the East received all-time record snows.

The NAO is discussed in more detail in chapter 11.

Location of Cold and Warm Pools

In a paper in the *Journal of Climate* in August 1998, Noel and Changnon studied storm track frequency versus factors such as the strength of El Niño and La Niña and subtle differences in the position of the warm and cold ocean pools. The results of their study showed that there were different flavors of El Niño and La Niña. We noted earlier in this chapter how the strength of an El Niño affected temperature patterns and the probability of heavy snows. Noel and Changnon's work suggests that the position had a significant effect

Figure 4.8.   The storm tracks in La Niña winters when the cold water plume extended well west of the dateline. In these La Niñas, storms came ashore further south on the west coast.
Courtesy of Intellicast.com from data by Noel and Changnon (1998)

on storm tracks. In La Niñas, for example, when cold water is confined more to the eastern Tropical Pacific, storms come ashore more frequently in the Pacific Northwest (Figure 4.7). When the cold water is situated further west into the central Tropical Pacific, the storms come ashore further south into California (Figure 4.8).

La Niña and Severe Weather Seasons

A look back at tornado climatology over the years reveals a tendency for the tornado frequency to be greater in La Niña years. Indeed, the weather patterns associated with La Niña are conducive to tornado outbreaks. In winters and early springs of La Niña winters, the jet stream tends to dip south of normal in the central and western states and rise to the north across the southeast (Figure 4.9). Storms come out of the southwest, redevelop over the Southern Plains, and head up toward the Ohio Valley or Great Lakes. They draw on the Gulf

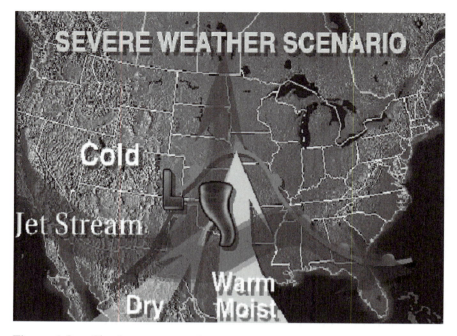

Figure 4.9.   Classic severe weather scenario occurs frequently in La Niña late winter and spring. The jet stream tends to dip down in the western United States and then up on the eastrn states. Storms can come out of the Rockies and feed on the warm moist air from the Gulf of Mexico. The strong overlapping jet streams can help cause rotation in developing thunderstorms and lead to large outbreaks of tornadoes.
Source: Intellicast—http://www.intellicast.com/DrDewpoint/Library/EL/
Courtesy of Intellicast.com

of Mexico's warmth and moisture and produce heavy rains and thunderstorms, and often severe weather in the lower Mississippi Valley and Gulf States. During the spring, the threat gradually shifts north into the western Tennessee and Ohio Valley Regions.

Bove and O'Brien, in their 1999 study, "Impacts of ENSO on United States Tornado Activity," found that during the winter and spring months, the south central states (Southern Plains to Louisiana and Arkansas and as far north as Iowa) shows significant decreases in tornado activity during El Niño. They found, on the other hand, that the Deep South and Ohio Valley experience a statistically significant increase during La Niña years.

Knowles and Pielke (1993) found that El Niño produced weaker tornadoes with shorter damage paths and fewer major outbreaks.

Table 4.2
Record Tornado Months and ENSO State

	January	February	March	April	May	June
Record #	216	83	180	267	335	311
Year	1999	1971	1976	1974	1991	1982
ENSO	La Niña	La Niña	La Niña	La Niña	Non-ENSO	El Niño
	July	**August**	**September**	**October**	**November**	**December**
Record #	242	126	139	100	146	96
Year	1993	1994	1967	1997	1992	1982
ENSO	El Niño	El Niño	La Niña	El Niño	El Niño	El Niño

Source: Data since 1950 from the NWS Storm Prediction Site http://www.nssl.noaa.gov/
~spc/products/svrstats.html. Courtesy of NWS

They also found that La Niñas are associated with stronger tornadoes that remain on the ground longer and tend to create more families of 40 or more tornadoes within a storm.

Ropelewski and Halpert (1986) noted that the temperature gradient across the United States tends to be less during El Niño (which is less conducive to tornadoes), while the gradients are enhanced in La Niña (creating a more favorable environment for tornadoes).

The La Niña Signal in Tornado Climatology

An examination of the record for most monthly tornadoes reveals that for the winter and early spring, the records are all La Niña years. Tornadoes in La Niña years tend to be more deadly as well. In fact, 7 of the 10 deadliest tornado years were classified as La Niña years (see Table 4.2).

Some Notable Severe Weather Outbreaks in La Niña Years

La Niña late winters and springs historically have brought some of the worst tornado outbreaks. One of the worst tornado outbreaks ever recorded occurred in April 1974, after a strong La Niña (Figure 4.10).

On April 3–4, 1974, 148 tornadoes touched down in 11 states and two Canadian provinces, killing 309 people and injuring 5,500. Included in the many killer tornadoes on that fateful day was the Xenia, Ohio storm that totally devastated that town (Figure 4.11). There

Figure 4.10.   Infared satellite image on April 3, 1974, the day of the Super Outbreak of 148 tornadoes. Note the multiple squall lines from the Great Lakes to the Gulf.
Source: NOAA NCDC web site—http://www.ncdc.noaa.gov/ol/climate/extremes/1999/april/sat3.gif
Courtesy of NOAA

have been just 50 tornadoes classified as F5 since 1950. On April 3, 1974, seven F5 storms occurred in Alabama, Kentucky, and Ohio.

Another super outbreak of tornadoes occurred on Palm Sunday in 1965 after a relatively weak La Niña. On that April 11, tornadoes killed 137 in Indiana, 62 in Ohio, and 53 in Michigan.

On April 21, 1967 (a La Niña year), 59 people died in an outbreak in Illinois. In the La Niña year of 1971, an outbreak of tornadoes on February 21 killed 118 in the Mississippi Delta region. In the La Niña spring of 1985, an outbreak in Pennsylvania killed 65 people.

A winter storm during January 21–24, 1999 produced more than 106 tornadoes, by far the largest winter season (December through February) outbreak on record (since 1950). This helped boost the January 1999 total to 216 tornadoes. This total was the largest monthly total for January or, for that matter, any winter month. The prior record for January was 52 tornadoes in 1975. The prior most

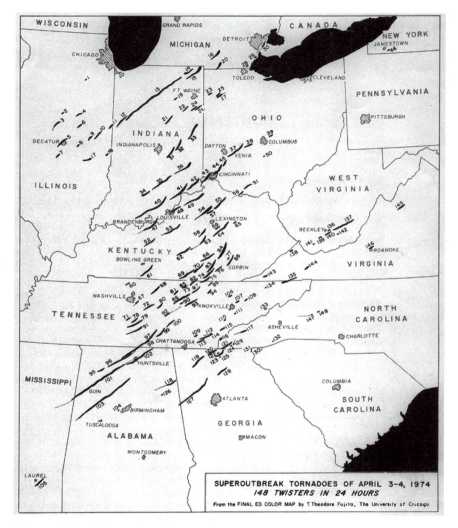

Figure 4.11.   Tornadoes during the Super Outbreak of April 3–4, 1974.
Source: National Climate Data Center (NCDC) Climate Watch 1999 http://
www.ncdc.noaa.gov/img/climate/extremes/1999/april/map.jpg
Courtesy of National Climate Data Center (NCDC)

active winter month was February 1971 (a La Niña year) when 83 tornadoes were confirmed. It even surpassed the March (the start of the tornado season) record of 180 tornadoes in 1976 (a La Niña year). Also on May 3, 1999, a violent storm system with 76 reported tornadoes raked central Oklahoma and the Wichita, Kansas areas, killing 44 people.

Chapter 5

ENSO and Summer Weather Patterns

DROUGHT

In the United States, drought produces average annual economic losses of between $6 and $8 billion, which is greater than floods ($2.4 billion) and hurricanes ($1.2 to 4.8 billion). ENSO is one of the factors that determines where drought is most likely to occur in a given year or season.

La Niña and Drought

In La Niña years, the regions throughout the tropics that are dry during El Niño years turn wet and the risk of flooding and disease often replaces drought and famine as the biggest concerns. However, areas that were wet during El Niño years now typically become dry. One of the driest areas is the southeastern and far southwestern United States. The storm track in La Niña years is most often across the northern United States. This brings heavy snows and rains to the north, but leaves the southernmost states with rainfall deficits. In the spring, as the sun returns, these areas are subject to brush, grass and forest fires. In stronger La Niñas, the winter drought can persist into the summer and become an agricultural drought.

In general, a ridge of high pressure at the surface and aloft tends to be stronger than normal near the central states during the La Niña summer. In strong La Niñas, this ridge is very strong and persistent and centered further north. When they become well established, these heat-wave ridges block weather systems from passing and bringing relief from the heat and needed rainfall.

In weaker La Niña events, the ridge is not as strong, moves around more, and permits weather systems to move and bring weather changes. The ridge results in a northwest upper level flow across the north central and northeastern states with more of a trough in the east.

Because of these differences, there is a large variance in summer temperature and precipitation patterns in strong versus weak La Niñas. Strong La Niñas are characterized in the mean by above to much above normal temperatures underneath the sprawling ridge across the north, while normal to cooler than normal temperatures prevail in the northwest and across the south. Potentially serious drought conditions often prevail across the central states into the Ohio and Tennessee Valley to the east coast (see Plate 3). Tropical moisture abounds to the south and above normal precipitation often occurs near the Gulf and in the Atlantic Coast states.

In weak La Niñas, a very different result often occurs. Temperatures average below normal in most of the central and eastern states. It is often drier than normal in the north central states, but with cooler than normal temperatures, the dryness has less of an impact. Wetter than normal conditions are typically found from the Gulf States up the Appalachians to the northeast (see Plate 4).

The central and southeastern United States often experience winter and spring drought during La Niña years. This pattern may play into the subsequent summer conditions.

A dry soil has a greater influence on the weather above during the warmest months. A dry soil over a large area leads to less evaporation, which in turn, leads to lower humidity and less cloudiness and precipitation. With more sunshine, temperatures rise more during the day and, as temperatures rise, humidities fall. In the long days of early summer, a heat wave can develop and over time grow in strength and area extent.

Over time, the atmosphere expands as it heats, creating a dome-like "heat ridge." Like a rock in a stream, this mountain of hot air diverts rain-bearing systems around it. Through the combined effects of reduced available moisture, increased temperatures and the blocking effects of these heat ridges, summer droughts are often self-perpetuating. Meanwhile, areas all along the edge of the hot, dry region can actually get flooding rains and severe weather. The concentration of stormy weather around the edges of a drought ridge is called the "ring-of-fire" effect. The strong thunderstorms feed off the extreme heat in the ridge, and on the enhanced jet stream there. This enhanced flow field is similar to what you find around a rock in a stream.

In strong La Niñas, the winter and spring dryness can be significant, and the lack of soil moisture can help make the heat ridge stronger and more persistent. It should be noted that La Niña heat waves and droughts can be very expensive. The 1988 heat wave and drought in central and eastern United States produced very severe economic losses amounting to an estimated $40 billion ($56 billion when adjusted to 1998 dollars) and an estimated 5,000 to 10,000 deaths (includes heat stress-related deaths) (Stevens 1989).

From the fall of 1995 through the summer of 1996, severe heat and drought in Texas and Oklahoma produced estimated losses of $5 billion. In the La Niña summer of 1999, the eastern drought and heat wave produced more than $1 billion in damage, and an estimated 256 deaths. The following winter and spring, parts of the midwest, central plains, and southern United States experienced severe drought. Arizona had the second driest October through April period on record. Central and southern Louisiana had rainfall deficits of 29 inches, Georgia deficits of 20 inches, and South Carolina deficits of 18 inches. Further north, even after heavy May rains, parts of Missouri were 11 inches below normal rainfall for the year. In New Mexico, 250,000 acres of drought-parched land were scorched by fire.

Effects of ENSO on Tropical Storms and Hurricanes

The El Niño and La Niña have effects on both the Atlantic and eastern Pacific tropical seasons. El Niños enhance the activity in the eastern Pacific and tend to suppress activity in the Atlantic. La Niñas have the opposite effect. They suppress activity in the eastern Pacific, and enhance it in the Atlantic hurricane season. The causes are both direct and indirect.

Bove et al. (1998) found that during the period 1900 to 1997, the probability of two or more hurricanes making landfall in the United States in El Niño years was 28 percent. The probability increased to 48 percent during neutral years, and 66 percent during La Niña summers.

El Niños are warm events with weakened subtropical high pressure centers and weaker than normal upwelling. Usually, the result is that waters are warmer than normal not only in the El Niño equatorial zones but areas north and south along the west coasts of North, Central, and South America. In La Niña years, the upwelling is enhanced and waters in these same areas tend to be cooler than normal.

Tropical storms and hurricanes that develop in the eastern Tropical Pacific usually form off the south and southwest coast of Mexico. Tropical storms and hurricanes feed off the heat and moisture given off by warm ocean water. Tropical storms are much more likely to develop when water temperatures reach more than 28°C or 80°F. This is evident both in computer models of these systems and in actual observations in nature.

In El Niños, the warmer water in the eastern Tropical Pacific helps more and stronger systems to develop over a longer season. In La Niña, the cooler than normal water suppresses hurricane formation, and seasons are thereby less active in the eastern Tropical Pacific. In the Atlantic, the indirect effects are most important. The El Niño tends to energize the high-level westerlies across the Caribbean and Tropical Atlantic. These westerly winds "shear" off the tops of thunderstorms in disturbances that are attempting to organize into tropical storms. As a result, more of these disturbances never make it to tropical storm or hurricane strength. In La Niña hurricane seasons, the high-level westerly winds are much weaker and the shearing is, typically, much less. As a result, more of the disturbances organize into tropical systems and hurricanes. These storms tend to develop sooner (often as the disturbances move off of Africa) and have more time to intensify. Because of this, more storms develop and are more likely to become intense (Category 4 or 5 hurricanes on the Saffir Simpson Scale).

Thus, in El Niño years, the west coast of Central America and Mexico and sometimes the southwestern United States are vulnerable to assault from more and stronger tropical systems.

The storms can even travel north into the normally arid southwest deserts or travel west and affect places such as Hawaii. The threat along the United States east coast is less. In La Niña years on the other hand, greater than normal hurricane activity threatens the east coast of the United States, the Caribbean, and the east coasts of Central America and Mexico. In La Niña years, the risk along the west coast of Mexico and Central America is reduced.

Some Examples of El Niño Tropical Storms in California

As indicated earlier, the eastern Tropical Pacific hurricane season is enhanced in El Niño years and the storms may affect the southwestern United States. The following examples of El Niño storms affecting the region was compiled by the National Weather Service in Oxnard, California:

July 1902: Remnants of a tropical cyclone that made landfall in southern Baja California produced rainfall of up to 2 inches in the deserts and southern mountains of southern California on July 20 and 21. This occurred during the strong El Niño of 1901–1902.

August 1906: A tropical cyclone tracked north/northwestward across the Gulf of California into the southwestern states generating rainfall of up to 5 inches in the deserts and southern mountains of southern California on August 18 and 19. This occurred during the El Niño of 1905–1906.

August 1915: Remnants of a tropical cyclone moved northward across northern Baja California into the deserts of southern California with rainfall of 1 inch at Riverside on August 26. This occurred during the strong El Niño of 1914–1915.

September 1918: Remnants of a tropical cyclone tracking to the north/northwest off the coast of Baja California and Southern California generated rainfall 7 inches in the mountains of northern California, but only light amounts for coastal areas of southern California on September 11 and 12. This occurred during the El Niño of 1918–1919.

September 1932: A tropical cyclone tracked north/northwestward across the Gulf of California. The remnants generated rainfall of up to 7 inches in the mountains and deserts of southern California over a 4-day period ending on October 1. Rainfall of 4.38 inches at Tehachapi in 7 hours on September 30 caused flash floods at Auga Caliente and Tehachapi creeks, resulting in 15 deaths. This occurred during the El Niño of 1932–1933.

September 1939: The first of four storms to affect southern California that month. The remnants of a hurricane tracked northeastward across northern Baja California into southwest Arizona generating rainfall of up to 7 inches in the southern mountains and southern and eastern deserts of southern California on September 4–7, with the heaviest rain on September 5 and 6. Blythe received more rain than would normally fall in 1 year and Imperial received more rain than would normally fall in 2 years. This occurred during the El Niño of 1938–1939.

September 1939: The remnants of a second tropical cyclone tracked northeastward across northern Baja California into southwest Arizona. Moisture from this tropical cyclone interacted with an upper trough to the north generating rainfall of up to four inches in the deserts and central and southern mountains of southern California on September 11 and 12. This occurred during the El Niño of 1938–1939.

September 1939: A tropical cyclone moving northwestward, just off the west coast of Mexico . . . moved into southern Baja California and dissipated. The moisture from this tropical cyclone generated rainfall of up to 3 inches in the deserts and central and southern mountains of southern California on September 19–21. This occurred during the El Niño of 1938–1939.

September 1939: Near the end of the month, a tropical cyclone moving to the northeast moved onshore at Long Beach at tropical storm strength with sustained winds of 50 mph. This is the only known eastern Pacific tropical cyclone to move onshore into southern California at tropical storm strength. Rainfall of 5 inches in the Los Angeles basin and 6 to 12 inches in the surrounding mountains occurred on September 25. The heaviest rain in the deserts occurred the day before with 6.45 inches of rain at Indio in a 6-hour period on September 24. This occurred during the El Niño of 1938–1939.

September 1941: Moisture from a north/northwestward moving hurricane that slammed into southern Baja California, generated rainfall of up to 1 inch in the southern mountains and deserts of southern California. This occurred during the strong El Niño of 1941–1942.

September 1946: A tropical storm moved northward into northern Baja California and dissipated with rainfall of up to 4 inches in the southern mountains of southern California on September 30 and exceeding 4 inches in the central mountains of southern California on October 1. This occurred during the El Niño of 1946–1947.

August 1951: A hurricane, moving north/northwestward just off the west coast of Baja California moved into northern Baja California and dissipated. Moisture from this tropical cyclone resulted in rainfall of 2 to 5 inches in the southern mountains and deserts of southern California on August 27–29. Many roads were washed out in the Imperial Valley, but otherwise no major damage occurred in southern California. This occurred during the El Niño of 1951–1952.

September 1952: A west/northwestward moving tropical storm southwest of Baja California dissipated. Moisture from this storm resulted in rainfall of up to 2 inches in the deserts and central and southern mountains of southern California on September 19–21 with most falling on September 19. This occurred during the El Niño of 1951–1952.

July 1954: A northward moving hurricane made landfall in central Baja California with the remnants moving into Arizona. Rainfall of up to 2 inches occurred in the deserts and southern mountains of southern California on July 17–19. This occurred during the El Niño of 1953–1954.

July 1958: Moisture from a west/northwestward moving tropical storm, which dissipated west of central Baja California, generated up to 2 inches of rainfall in the deserts and southern mountains of southern California on July 28 and 29. This occurred during the El Niño of 1957–1958.

September 1963: A northeastward moving tropical storm, Katherine, made landfall in northern Baja California. Rainfall of up to 7 inches in the central and southern mountains of southern California fell on September 17–19. This occurred during the El Niño of 1963–1964.

September 1965: A north/northwestward moving Hurricane Emily dissipated just off the west coast of central Baja California with spotty rainfall amounts up to 1 inch in the mountains of southern California on September 4 and 5. This occurred during the El Niño of 1965–1966.

September 1972: Hurricane Hyacinth moved as far west as 125°W longitude before recurving to the northeast. The remnants made landfall between Los Angeles and San Diego on September 3 with winds of 25 mph and rainfall of up to 1 inch in the central and southern mountains of southern California. This tropical cyclone holds the distinction of traveling the farthest west before recurving and making landfall in southern California. This occurred during the El Niño of 1972–1973.

October 1972: Hurricane Joanne recurved making landfall in northern Baja California and maintaining tropical storm strength into Arizona and generating rainfall up to 2 inches in the southeast deserts of southern California on October 6. This occurred during the strong El Niño of 1972–1973.

September 1976: North/northwestward moving Hurricane Kathleen made landfall in northern Baja California with the remnants moving into southern California. Hurricane Kathleen brought to the southwest the highest sustained winds ever associated with an eastern Pacific tropical cyclone with sustained winds of 57 mph at Yuma on September 10. Six to 12 inches of rainfall was observed in the central and southern mountains of southern California on September 10 and 11. Most of Ocotillo, California was destroyed by flooding, and three persons drowned. This occurred during the El Niño of 1976–1977.

August 1977: Hurricane Doreen tracked north/northwestward along the West Coast of Baja California, dissipating over the southern California coastal waters. Most areas of southern California from Los Angeles south received at least 2 inches of rainfall with up to 8 inches in the mountains. Flooding was widespread with extensive crop damage. This occurred during the El Niño of 1977–1978.

October 1977: Hurricane Heather recurved with the remnants tracking across northern Baja California into Arizona. There was rainfall up to 2 inches in the southern mountains and deserts of southern California on October 6 and 7. This occurred during the El Niño of 1977–1978.

September 1978: Hurricane Norman recurved with the remnants tracking into Southern California from the south-southwest. Rainfall was most intense on September 5 and 6 with amounts exceeding 3 inches in the mountains of southern California. This occurred during the El Niño of 1977–1978.

September 1982: Remnants of Hurricane Norman tracked northeastward across northern Baja California into Arizona with scattered rainfall amounts up to 1 inch in the southern mountains and deserts of southern California on September 17 and 18. This occurred during the strong El Niño of 1982–1983.

September 1982: The remnants of Hurricane Olivia recurved northeastward across southern California with rainfall up to 4 inches in the mountains of southern California from September 24–26. This occurred during the strong El Niño of 1982–1983.

September 1983: Northward moving Hurricane Manuel dissipated off the west coast of northern Baja California with up to 3 inches of rainfall in the southern mountains and deserts of southern California on September 20 and 21. This occurred during the strong El Niño of 1982–1983.

October 1983:The remnants of northward moving Hurricane Priscella tracked across southern California with only scattered light rainfall on October 7. This occurred during the strong El Niño of 1982–1983.

August 1997: The remnants of Hurricane Ignacio tracked northward moving inland in central California with gale force winds over portions of the southern California coastal waters. This occurred during the strong El Niño of 1997–1998.

September 1997: Hurricane Linda developed in the eastern Pacific. Although the storm did not have a major effect on southern California, it was notable for its strength. With winds of 185 mph, it prompted some meteorologists to propose a new category, a Category 6 on the Saffir-Simpson scale for Super Hurricanes.

For the east coast, hurricanes are a much greater threat during La Niña summers. Overall, the odds that the United States will experience major damage from an Atlantic tropical system is much greater in La Niña years than El Niño years because there are more storms, and the storms tend to be more intense. The average damage per storm of El Niño years is $800 million versus $1.6 billion in La Niña years (Pielke & Landsea, 1999). An analysis by Florida State students found that the chances for two or more landfalling hurricanes in El Niño years is 25 percent whereas in La Niña it is 75 percent. This does not mean El Niño years are hurricane-free or always impact-free. Hurricane Betsy, in 1965, resulted in more than $13 billion in damages. Hurricane Agnes, in 1972, produced more than $11 billion and Andrew, in 1992, more than $20 billion.

During the 1999 La Niña, the one–two punch of Hurricanes Dennis and Floyd brought up to three feet of rain to the eastern third of North Carolina, a region that had been suffering from drought. The heavy rains produced floods, which created havoc in the large estuary that separates the Outer Banks from the mainland. The floodwaters picked up decomposing vegetation, farm and lawn fertilizer, raw sewage, hog waste from containment farms, and topsoil. This formed the seeds for what could have been an ecological time bomb. The concern was that in the spring as temperatures warmed there would be large blooms of algae called phytoplankton. As these algae died, bacteria would feed on them and use up tremendous amounts of oxygen from the water, which could have been devastating to fish and other creatures.

A review of hurricanes during La Niña summers reveals an increase in frequency and some real blockbusters. In the dozen La Niña summers since 1950, there were 135 tropical storms (11.7 per year versus an average of 9.3) and 82 hurricanes (6.8 per year versus an average of 5.8).

Ten Famous La Niña Hurricanes

The Great Hurricane of 1938 (Category 5)

This great hurricane took forecasters and residents of the northeast by surprise. Residents of the region had come to believe they were safe from strong hurricanes. It had been decades since a major hurricane had struck New England and 150 years since one cut a swarth through the interior. Very few residents of the region were "hurricane conscious" and virtually no one was prepared for what was about to happen.

Late on September 20, a tropical storm was reported in the eastern Bahamas. In that position, storms are usually 3 days away from threatening the northeast coast. Charlie Pierce, a junior forecaster in the U.S. Weather Bureau, predicted the storm would come north, but was overruled by the chief forecaster. The storm was instead forecast to behave like most storms had been doing and recurve east of the Carolinas and out into the open Atlantic.

However, this storm was moving into a region with a deepening trough of low pressure, which captured the storm and accelerated north up the coast. It was estimated to be moving at more than 60 mph when it surprised Long Island and southern New England on the afternoon of September 21 before any warnings could be issued. After crossing Long Island, the hurricane made landfall in Connecticut, putting Rhode Island on the more dangerous, right side of the storm center.

The strongest winds were estimated at 160 mph (a Category 5 storm). Blue Hill Observatory in Massachusetts had sustained winds of 121 mph with gusts to 186 mph.

Along the coast of both Long Island and southeastern New England, a storm surge as high as 25 feet topped by wind waves as high

Figure 5.1. Waves striking seawall look like erupting geysers. New England coast, 1938.
Source: Image wea00412. Photo by NOAA Photo Library—http://www.photolib.
noaa.gov/historic/nws/images/wea00412.jpg
Courtesy of NOAA archives

as 30 feet inundated coastal communities. Damage from wind and flooding, due to heavy rains up to 17 inches, was widespread in all but northernmost New England (see Figure 5.1).

In Providence, Rhode Island's largest and capitol city, massive flooding occurred as Narragansett Bay rose to 13 feet above mean low water. Three hundred eighty people in Rhode Island died and 564 deaths overall were attributed to the hurricane. Many believe the death toll may have been closer to seven hundred. Another 1,700 were injured. The "Great Hurricane of '38" remains the fourth deadliest hurricane in U.S. history.

A total of 16,740 structures were destroyed and many more damaged. Some 16,000 families were displaced or left homeless. The toll for the mariners could only be described as catastrophic. More than

2,600 boats were lost or destroyed beyond repair and another 3,300 damaged.

An estimated 275 million trees were downed along with 20,000 miles of electric power and telephone lines. The wind-driven rains carried dissolved ocean salt, which damaged vegetation 50 to 100 miles inland. There were reports of a sea-salt residue on windows in Montpelier, Vermont.

During the rainfall that preceded and accompanied the hurricane, an average of eleven inches of rain fell over a 10,000 square mile area. Four days of rain culminating in the hurricane downpours left 10 to 17 inches in the Connecticut River Valley, resulting in some of the worst flooding ever recorded there. The wave of flooding inflicted major damage from New York and Connecticut to Massachusetts and Vermont.

In 1990 dollars, the "Great Hurricane of '38" left in its wake approximately $3.6 billion damage throughout New York and New England. Christopher Landsea, a meteorologist at the Hurricane Research Division and Roger Pielke, a social scientist at the National Center of Atmospheric Research, looked at the most destructive hurricanes and estimated the cost if they were to hit today. The "Great Hurricane of '38," they found, would be the sixth costliest of all time. In 1998 dollars, their study estimates that a repeat of the 1938 storm would produce $18 billion in damages.

1954: Hurricane Carol (Category 3)

The most destructive hurricane since 1938 smashed through eastern Long Island and southeastern Connecticut on the morning of August 31, 1954. Like the "Great Hurricane of '38," Carol accelerated north from the Bahamas passing just east of Cape Hatteras, North Carolina on the evening of August 30.

Carol made landfall on eastern Long Island and southeastern Connecticut about twelve hours later with sustained winds of 80 to 100 mph. Block Island, Rhode Island recorded the strongest winds ever there with gusts to 135 mph. Trees and power lines were downed and crops were devastated.

Carol came ashore just after high tide causing widespread flooding. Storm surges ranged up to 15 feet east of New London, Connecticut. On Narragansett Bay, a storm surge of 14.4 feet exceeded that of the storm of 1938 however, because it occurred after high tide, the storm tide was less.

Coastal communities were hard hit from New London, Groton, and Mystic in Connecticut to Westerly to Narragansett in Rhode Island. Downtown Providence was flooded under 12 feet of water nearly the same as in 1938. Up to 6 inches of rain occurred from near the point of landfall up to northeastern Massachusetts.

The storm destroyed nearly 4,000 homes, 3,500 automobiles, and 3,000 boats. Eastern Connecticut, all of Rhode Island, and much of eastern Massachusetts were without power and 95 percent of the phone service was interrupted. The damage adjusted to 1996 dollars was $2.7 billion.

1954: Hurricane Edna (Category 3)

Before the region devastated by Hurricane Carol could fully recover, Hurricane Edna began a similar track up the East Coast on September 10, 1954. But then Edna veered about one hundred miles further east making landfall near Martha's Vineyard and Nantucket before moving across eastern Cape Cod.

Edna brought hurricane winds of 75 to 95 mph to all of eastern Massachusetts and Rhode Island with a peak wind gust of 120 mph on Martha's Vineyard. Electrical power was disrupted in parts of Rhode Island and eastern Massachusetts including Cape Cod and the Islands. The storm surge reached 6 feet across Martha's Vineyard, Nantucket and Cape Cod producing serious flooding. The boating community was especially hard hit. More heavy rains brought the two hurricanes' (Carol and Edna) rainfall totals to as much as eleven inches from southeastern Connecticut and Rhode Island to northeast Massachusetts, forcing rivers several feet above flood stage. The death toll reached 21.

1954: Hurricane Hazel (Category 4)

Just when the residents of the east were thinking the worst had past, the third major landfalling hurricane, Hazel, came ashore in the Carolinas on October 15, 1954. In North Carolina, the storm surge reached eighteen feet at Calabash. It was made worse by occurring at the exact time of the highest lunar tide of the year, the full moon of October. This may have boosted the highest tide by several feet.

Estimates of 150 mph winds were reported from Holden Beach, Calabash, and Little River Inlet. Winds were estimated at 125 mph at Wrightsville Beach and 140 mph at Oak Island. The extent of the hurricane-force winds and the amount of time and distance these winds persisted after landfall were quite unusual. Fayetteville reported gusts of 110 mph with estimates of gusts to 120 mph at Goldsboro, Kinston, and Faison. Further away from the storm center, winds reached 100 mph in numerous locations in Virginia, Maryland, Pennsylvania, New Jersey, and New York.

The damage in North Carolina was likened to the battlefields of Europe after World War II with incredible tree damage, 15,000 homes and other buildings destroyed, and 39,000 others damaged. The damage (in 1996 dollars) was nearly $1.7 billion.

1955: Hurricanes Connie (Category 4) and Diane (Category 3)

The battering of the east coast that marked the summer and fall of 1954 continued in the summer and fall of 1955. Hurricanes Connie and Diane broke the heat of that hot summer with back-to-back thrashings of the middle-Atlantic and northeast states.

Connie moved ashore first near Cape Lookout in North Carolina with 145 mph winds on August 12. From there, the storm drifted northwest to the eastern Great Lakes where it dissipated.

Connie produced heavy rains as far away from the center as southern New York and New England where 4 to 6 inches of rain fell. This saturated the ground and brought rivers and reservoir levels to above normal. That set the stage for disastrous flooding with Diane less than a week later. Diane came ashore in North Carolina with 120 mph winds, however Diane recurved northeast across Virginia, New Jersey, and then just south of Long Island.

Diane brought heavy rains and flooding to the mid-Atlantic but reserved its worst damage for New England. It produced up to twenty inches of rain over a 2-day period. One hundred eighty people died from the storm. Connecticut was especially hard hit. Seventy-seven lives were lost in Connecticut and the damage there was $2.5 billion (1996 dollars). Massachusetts was also very hard hit. The Westfield River exceeded its previous flood stage by nearly five feet.

Overall in New England, more than two hundred dams suffered partial to total failure, especially in the Thames and Blackstone headwaters.

The damage from Diane (with preconditioning from Connie) was estimated at about $4.8 billion (1996 dollars), putting it into the top 10 of the costliest hurricanes on record for the United States.

1988: Hurricane Gilbert (Category 5)

Hurricane Gilbert began as a tropical wave moving off the coast of Africa on September 3, 1988. The wave continued westward in the trade wind belt and by September 8 was approaching the Lesser Antilles, at which time it was classified a tropical depression. On September 9, satellites indicated better organization and reconnaissance aircraft found tropical storm winds and the storm was named.

Gilbert strengthened rapidly to Category 1 hurricane status by September 10 and Category 2 with winds of 96 to 110 mph just twelve hours later. It passed just south of the Dominican Republic and then struck Jamaica as a Category 3 storm with winds of 111 to 130 mph. It continued to strengthen as it passed Jamaica to a Category 4 storm when it passed south of the Grand Cayman Islands on September 13, with winds of 131 to 154 mph.

Shortly after it passed Grand Cayman, Gilbert strengthened to a Category 5 storm with winds more than 155 mph. By 6 p.m. on September 13, Gilbert had reached the lowest pressure ever recorded in the Western Hemisphere—888 mb (26.22 inches Hg). It is likely the storm actually was stronger (at least 885 mb) in between observations.

Gilbert had a small eye (characteristic of very strong storms) but its overall size was huge—at one point covering the entire western half of the Caribbean, Central America, and southeastern Gulf of Mexico.

The storm moved ashore in the Yucatan as a Category 5 storm, the first to do so since Camille in 1969. A 20-foot storm surge accompanied Gilbert onshore. Cancun and other resorts were very hard hit.

A little more than a month later, another Caribbean storm, Joan, a Category 4 storm, struck Nicaragua on October 22 with sustained winds of 135 mph. Joan and Gilbert combined to produce more than $5 billion in damage and more than 500 deaths through the Caribbean, Mexico, and Central America.

1989: Hurricane Hugo (Category 5)

Hurricane Hugo began as a tropical wave off of Africa on September 10, 1989. In its 2-week lifetime it became a Category 5 storm before battering the Virgin Islands and Puerto Rico on September 17 and 18, weakened to a Category 2 then re-intensified to a Category 4

Figure 5.2.   Large oak trees over 100 years old came down all over Charleston after passage of Hurricane Hugo.
Source: Image wea00477. Photo by National Hurricane Center, NOAA Photo Library.
Courtesy of NOAA archives

storm before swirling through South and North Carolina on September 21 and 22. The damage in the Virgin Islands and Puerto Rico was described as tremendous. The destruction in Puerto Rico caused mass riots, looting, and many post-storm deaths.

The storm punished South Carolina to the tune of $5.9 billion. Twenty-nine people died in South Carolina and 50,000 were left homeless. Forty-two of the 46 South Carolina counties were declared disaster areas.

In South Carolina, Hugo produced wind gusts as high as 179 mph at Bulls Island, just northeast of Charleston, and a storm tide as high as 19.8 feet at Romain Retreat (Figure 5.2). The look of beautiful Charleston was changed overnight. More than 10,000 trees were downed and many thousands more damaged by the storm. Many of the trees were old and majestic and provided a special kind of beauty to this "Southern Belle" of a city. Trees were downed as far north as Charlotte in North Carolina. The timber industry suffered more than $1 billion in losses from Hugo.

1996: Hurricane Fran (Category 3)

Hurricane Fran was the powerful second part of a two-stage tropical assault on North Carolina during the summer of 1996. Hurricane Bertha came ashore in July in North Carolina as a Category 2 hurricane. Fran arrived as a Category 3 storm less than two months later.

Fran began as a tropical depression on August 24. From August 29 to 30, Fran passed north of the Lesser Antilles as a Category 1 hurricane, weakened to a tropical storm, then regenerated back to a hurricane while travelling toward the Bahamas.

Fran intensified to a Category 3 storm with winds of 115 mph before making landfall east of Cape Fear in North Carolina on September 5, 1996. The storm's twelve-foot storm surge carried away a police station and town hall, temporarily housed in a double-wide trailer after Hurricane Bertha damaged the original offices in July. Fran caused extensive flooding in nearby Wrightsville Beach and other coastal communities. There was also considerable tree and building damage not only in North Carolina but also in Virginia and even parts of Maryland. At some point, 1.7 million customers in North Carolina and 400,000 in Virginia were without power.

Rains of up to fifteen inches produced heavy flooding across parts of interior North Carolina, Virginia, and West Virginia.

At least thirty-seven people died and overall damage exceeded $5 billion. The damage to homes and businesses was estimated as $2.3 billion; to public roads, buildings, utilities $1.1 billion; to agriculture $700,000; and timber about $1 billion.

1998: Hurricane Mitch (Category 5)

Very late in the 1998 hurricane season on October 26–27, Meandering Monster Mitch became the fourth strongest Atlantic Basin Hurricane ever with winds exceeding 180 mph. Hurricane Mitch was the strongest storm in the western Caribbean since Gilbert in 1988. Mitch stalled off the coast of Honduras from late on October 27 to the evening of October 29 before moving onshore. The storm dumped incredible rains on the mountains of Central America, causing floods and mudslides that were responsible for the death of between 10,000 and 12,000, making it one of the top five deadliest hurricanes on record in the Atlantic Basin.

Hurricane Mitch was another on the list of La Niña Atlantic/Caribbean powerhouses to affect this region. Some other very powerful

(Category 5) La Niña year storms affecting this region included Janet in 1955, Edith in 1971, and the memorable Gilbert in 1988. Other strong systems that did not reach Category 5 status but which caused deaths and significant damage in the region included Ella (Category 3) in 1970, Carmen (Category 4) in 1973, Joan (Category 4) in 1988, and Roxanne, a meandering Category 3 storm in 1995.

1999: Hurricane Floyd (Category 4)

Floyd was a large and intense Cape Verde hurricane that pounded the central and northern Bahama islands, seriously threatened Florida, struck the coast of North Carolina and moved up the United States East Coast into New England. It neared the threshold of Category 5 intensity on the Saffir/Simpson Hurricane Intensity Scale as it approached the Bahamas, and produced a flood disaster of immense proportions in the eastern United States, particularly in North Carolina.

Floyd slowly strengthened and became a hurricane by September 10, 1999, while centered about 200 nmi. east-northeast of the northern Leeward Islands. Hurricane Floyd made landfall near Cape Fear, North Carolina on September 16 as a Category 2 hurricane with estimated maximum winds near 100 mph.

Heavy rainfall preceded Floyd over the mid-Atlantic states due to a pre-existing frontal zone and the associated overrunning. Therefore, even though this tropical cyclone was moving fairly quickly, precipitation amounts were very large. Rainfall totals as high as 15 to 20 inches were recorded in portions of eastern North Carolina and Virginia. At Wilmington, North Carolina, the storm total of 19.06 inches included a twenty-four-hour record of 15.06 inches. Totals of twelve to fourteen inches were observed in Maryland, Delaware, and New Jersey. New records were set in Philadelphia for the most rain in a calendar day, 6.63 inches. In southeastern New York, rainfall totals were generally in the four- to seven-inch range but there was a report of 13 inches at Brewster. Totals of nearly 11 inches were measured in portions of New England.

There were 57 deaths that were directly attributable to Floyd, 56 in the United States, and one in Grand Bahama Island. Most of these deaths were due to drowning from freshwater flooding. Floyd was the deadliest hurricane in the United States since Agnes of 1972.

In the United States, the Property Claims Services Division of the Insurance Services Office reports that insured losses due to Floyd totaled $1.32 billion. Ordinarily this figure would be doubled to estimate the total damage. However, in comparison to most hurricane landfalls, in the case of Floyd there was an inordinately large amount of freshwater flood damage, which probably alters the two-to-one damage ratio. Therefore, total damage estimates ranged from $3 to more than $6 billion.

Chapter 6

Other Effects of ENSO

HUMAN HEALTH

We know that weather and climate affect how we feel, perform, and behave. Day-to-day weather changes can affect ailments such as arthritis, emphysema, or asthma. Climate extremes, including those associated with El Niño or La Niña, can directly or indirectly influence the incidence and spread of disease.

Dr. Paul R. Epstein, associate director of Harvard University Medical School's Center for Health and Global Environment, writes:

> Changes in temperature, precipitation, humidity, and storm patterns, often related to the El Niño-Southern Oscillation (ENSO) phenomenon, are associated with upsurges of water-borne diseases such as hepatitis, shigella dysentery, typhoid, and cholera; of vector-borne pathogens such as malaria, dengue, yellow fever, encephalitis, schistosomiasis, plague and hantavirus; and of agricultural pests such as rodents, insects, fungi, bacterium, and viruses. And because climate variability can be forecasted, the potential exists to predict the likelihood of outbreaks of infectious disease.

Ecosystem Disruptions

Climatic extremes effect animal, plant and human health by affording opportunistic species fresh terrain and generating new bursts of activity. Droughts encourage locusts and rodents, while floods foster fungi and mosquitoes. Fluctuations in climate, which alter the structure or function of ecosystems, can change the population dynamics of opportunistic pests and disease vectors, and of the predatory species that normally check their population growth. Owls, for example, help control rodent populations involved in Lyme disease and Hantaviruses. In the United States, deforestation in the Pacific northwest and prolonged drought in the southwest, both encourage the proliferation of rodents by damaging owl refuges.

Rodents are involved in the life cycle of many groups of diseases around the world. In the United States, a new disease, the Hantavirus (with a 60 percent mortality rate), emerged in the "Four Corners" area near the borders of Colorado, Nevada, New Mexico, and Utah, following an explosion of the deer mouse population. A 6-year drought by 1993 in the southwestern states devastated populations of owls, snakes, and other rodent predators, and was followed by heavy rains that increased food sources for rodents; in the absence of predators the well-nourished rodents flourished. The story is similar in southern Africa, where heavy rodent infestations closely followed the El Niño years of 1976, 1983, and 1993.

Since the 1960s, researchers in southern Asia have observed an association between algal blooms and upsurges of cholera. It is becoming increasingly clear that ENSO warm events are associated with upsurges of cholera, perhaps via the marine reservoir and/or the contamination of ground water accompanying ENSO-related flooding. Recent increases in coastal alga blooms and related cholera epidemics may be linked to climatic perturbations of ecosystems already stressed by pollution, habitat destruction, or the introduction of nonindigenous species.

ENSO FORECASTING AS A WEAPON AGAINST INFECTIOUS DISEASE

New developments in climate forecasting can provide the basis for a proactive approach to the spread of human diseases and agricultural pests, thereby mitigating or preventing outbreaks before they occur, saving scarce public health resources and, ultimately, saving lives. By integrating health surveillance with environmental and climatological monitoring, early warning systems for conditions conducive to disease outbreaks can be developed and produce timely, ecologically sound, and environmentally benign public health interventions. Climate forecasting can also be extremely useful in targeting scarce funding for surveillance and response, research and training, and emergency production of vaccines, drugs, and diagnostics in the United States and abroad.

Although climate forecasting cannot predict exactly where, when, or to what extent outbreaks will occur, current forecast capabilities, combined with better understanding of the links between climate and health, can be used in assessing the vulnerability of populations to outbreaks of infectious disease, and in determining the risk of epidemics. Even at this early stage, probabilistic climate forecasting can arm public health practitioners with a powerful tool for reducing the morbidity and mortality caused by outbreaks of infectious disease. In this case, as in others, chance favors the prepared mind.

EL NIÑO EFFECTS ON MARINE LIFE

El Niño has a major effect on some of the world's largest fishing grounds. The coastal waters off Peru and California rank among the top five fishing grounds in the world because of the upwelling of cold, nutrient-rich water. This is caused by the surface trade winds that push the surface water away from the land. Lower temperature water then rises up from beneath to replace it. The process is a slow one, ranging generally from 20 to 100 meters per month depending on the location, the season, and ENSO.

The upwelling brings up chemical nutrients, principally phosphates and nitrates that stimulate plant growth. These nutrients accumulate in the ocean depths from the debris of dead marine plants and animals. They are then carried up with the upwelling water toward the surface. There, together with sunlight, these nutrients help drive the photosynthesis in microscopic plants called phytoplankton that, in turn, provide food for billions of small-grazing animals, including minute crustaceans and a wide variety of small herbivores (plant-eaters). Anchovy larvae consume large quantities of this ocean life and are abundant when these upwelling-induced "bloom" conditions exist. Larger fish and squid then feed on the anchovies, as do ocean birds.

During the southern hemisphere summer (December to March), when the upwelling weakens and is more confined to near the shore, the anchovies confine themselves to the coldest waters near the shore. Because they tend to be concentrated at this time, they are easier prey for other fish, birds, and fishermen.

During El Niños, when the water, even along the coast, warms and the upwelling increases further, the life of the anchovy is especially disrupted. Due to the warmer water, warm-water species like the yellowfin tuna, dolphin, manta ray, and hammerhead shark, appear and feed on the anchovy. Even more of a threat, however, is the reduction in nutrients, which reduces the phytoplankton and the small animals that the anchovy depends on for food. In addition, the warmer water doesn't suit the anchovy, which prefer cooler water. Reduced in number, the anchovy scatter, no longer forming enormous schools. Sardines too abandoned the region, moving south to cooler, Chilean waters.

Other marine life suffers as these few important links in the food chain break. Birds that feed on the anchovy starve or fly away, abandoning their young and their nests. They travel over enormous stretches of ocean in a desperate search for food. Other species of fish, squid, sea turtles, and even small ocean mammals either migrate or die. The fur seal and sea lion populations are often decimated in strong El Niños. In 1982–1983, 25 percent of the adult populations and all of the pups died. In the El Niño of 1997–1998, sea lions on San Miguel Island off the coast of California were starving because the water was too warm to support the fish the sea lions normally feed on.

There are other domino effects. With the reduced fish catch, there is a corresponding reduction of the fishmeal produced and exported to other countries for feeding livestock and poultry. Other more expensive feed sources are required, which can mean an increase in poultry and livestock prices worldwide.

Even after the El Niño fades, it takes years for the animal populations to return to prior levels. The strong El Niño of 1971–1972, in which waters warmed to over 30°C along the coast of South America in February 1972, drastically decimated the anchovy population. The catch that year dropped more than 50 percent to levels lower than in a decade. It took a series of cold events to bring the levels back up.

The sudden and catastrophic destruction of plankton and fish and other marine life produces still other effects. Dead fish on beaches leads to the formation of hydrogen sulfide gas, which, in turn, can blacken the paint on ships.

Further north in the Pacific waters off North America, the salmon catch is reduced as the warm water related to El Niño pushes Pacific mackerel northward, where they eat young salmon and compete with adult salmon for food.

The warm waters flood the west coastal areas of California with exotic species such as barracuda and yellowtail. During this time, tuna can be spotted all the way north to Alaska. In the 1997–1998 El Niño, a 100-pound marlin, whose species normally prefers northern Mexico waters, was caught off Washington State.

Coral Reef Damage

In the western Pacific, the sudden drops in sea level due to an El Niño expose and destroy the upper levels of the very fragile coral reefs surrounding most islands. The coral reef is a very unique and rich ecosystem, which supports a vast array of animal and plant species.

The deadly combination of lowered sea levels and warm ocean temperatures resulted in significant damage to the coral reefs off Australia in the 1997–1998 El Niño. These reefs are home to a wide variety of marine life. Coral thrives at temperatures below 28°C (82°F). When water temperatures rise above 28°C however, a process known as coral bleaching kicks in, during which the coral expels the algae (microscopic organisms called Zooxanthallae) that thrive within the coral and are necessary for its survival. The coral and the algae have a special symbiotic relationship in which the algae supply oxygen and some organic compounds to their coral hosts in turn for their hospitality. When the algae is expelled, the coral polyps lose their pigmentation and appear transparent on the animal's white skeleton. If warm temperatures persist, other marine life that thrives in the coral is also affected, and ultimately the quality of fishing and recreation in the region also suffers.

In this same El Niño event of 1997–1998, in the eastern Pacific near the Galapagos Islands to the coast of Ecuador and Panama, even though the sea levels were above normal, extremely high ocean temperatures (over 30°C) resulted in serious damage to the sensitive coral. If coral reefs survive bleaching, they can recover when temperatures return once again to normal, although their recovery may take up to a year.

Chapter 7

How Scientists Study El Niño

RECONSTRUCTING HISTORICAL ENSO EVENTS

Current knowledge of how the atmosphere and oceans behave during ENSO events allows researchers to piece together a history of the ENSO status as far back as accurate and detailed observations exist. This provides history to approximately 1875. Proxy data (tree ring, coral, sediment records, South American fishery records, and ice core studies) has been used to evaluate the ENSO state further back in time.

By looking at these data sets in regions usually affected by ENSO, researchers can infer El Niño and La Niña events from the evidence of very wet or very dry years. The University of Arkansas Tree Ring Lab studied tree ring data from teakwood forests in Java and fir trees in Mexico and the American southwest that date back to around 1700. The thicker the tree rings, the more rain fell that year. In Mexico and the southwest, the thicker bands occur in El Niño years when rainfall is heaviest. In Java, El Niño years are drought years, whereas La Niña years bring the heaviest rains.

Coral reef beds have likewise been used to help piece together the ENSO history. The coral is an effective measure of both temperature and precipitation. When water temperatures rise, the small creatures incorporate less strontium into their skeletons than when the water is cool. Their oxygen content records change in salinity, which result from too much or too little precipitation. Ancient sediments in the deserts of Peru can also help establish the long-term ENSO history. The desert is one of the driest spots on earth except in El Niño years when heavy rains compact the surface dust into a layer of fine, reddish soil. Evidence from these sources suggest El Niños have been around for 2 million years or more.

Ice cores taken from mountain areas of the Andes in Peru likewise show the variations typical of El Niño and La Niña. The fishery records from Peru and Ecuador can also be used to identify significant events (both warm and cool). El Niños reduce the catch while the cool La Niñas often bring good catch years.

To help reconstruct history, researchers have employed using information from available historical writings, for example, the writings of Spanish colonists in settlements along the coasts of Peru and Ecuador dating back to the late 15th century. They have been used to identify at least some key events. Based on these sources, scientists have compiled the following list of possible very strong El Niño events: 1661, 1694–1695, 1782–1784, 1790–1793, 1844–1846, 1876–1878, 1899–1900, 1940–1941, 1982–1983, and 1997–1998.

CURRENT WAYS TO MEASURE ENSO

Because ENSO is a coupled atmospheric and oceanic phenomenon, conditions in both the atmosphere and ocean can be used to evaluate the ENSO state and provide a measure of the strength of an event. The three most commonly used measures of ENSO used today are the SOI, sea surface temperature anomalies (departure from normal) in the eastern and central Tropical Pacific, and the new Multivariate ENSO Index (MEI). The SOI uses atmospheric pressure; the sea surface temperature anomalies are obviously oceanic; whereas the MEI evaluates both atmospheric and oceanic parameters.

The SOI

The SOI is the oldest measure and was first designed by Walker in the 1920s and revised later in the 1950s by Willet and others. The SOI compares the pressures (relative to normal) at Darwin, Australia and the central Pacific Island of Tahiti. In El Niños, pressures are unusually high in the western Pacific (Darwin) and unusually low in the central Pacific (Tahiti). This causes a weakening or even a reversal of the easterly trade winds and a warming of the water in the eastern

Tropical Pacific. The SOI tends to be persistently negative during El Niños.

In La Niñas, the opposite occurs. Pressures are usually low in the western Pacific and high to the east. This causes a strengthening of the easterly trade winds and a cooling of waters in the eastern Tropical Pacific. The SOI tends to be persistently positive in La Niña (cold) events.

Surface observations have been used since before 1900 to compute historical SOI fluctuations and the related ENSO events.

Sea Surface Temperature Anomalies in the Tropical Pacific

If El Niños are warm ocean water events and La Niñas are cold ocean water events, it is not surprising that sea surface temperatures would be useful as a measure of ENSO. The region in the eastern Pacific, where the warm and cold plumes of the ENSO cycle is most apparent, stretches from along the northwest coast of South America west along the equator to beyond the International Dateline. This region has been divided into subregions and sea temperatures have been averaged in these regions and compared to normal. Historical sea surface temperature data has been analyzed in these regions to help identify ENSO events.

The eastern most regions are called NINO 1 and 2. They stretch from 80 to 90 degrees west longitude and from the equator to 10 degrees south latitude. Region 3 extends from 90 to 150 degrees west longitude and between 5 degrees north and 5 degrees south. Region 3.4 is defined as the region from 120 to 170 degrees west longitude and between 5 degrees north and 5 degrees south. Region 4 is farthest west, extending from 150 degrees west to 160 degrees east and again from 5 degrees north to 5 degrees south.

The most commonly used regions for ENSO monitoring are Regions 3 and 3.4. In El Niño events, sea surface temperature anomalies are greater than 0.5 degrees Celsius above normal. In the strongest events, the anomalies over the key regions exceed 2 degrees Celsius. In La Niña events, sea surface temperature anomalies are more than 0.5 degrees Celsius below normal. In the strongest events, temperature anomalies may exceed 2 degrees below normal.

The MEI

Research in the printed literature is somewhat inconsistent as to which years are ENSO years and how strong they were, depending on whether the researchers used the SOI or sea surface temperatures; and if sea surface temperatures was used, which region.

The Climate Diagnostics Center (Wolter, 1998) developed the MEI to provide a new comprehensive data set that incorporated multiple factors, including air temperatures, sea surface temperatures, sea-level pressure, surface wind, and cloudiness. The hope was that this reference data set would become the standard for all researchers to provide the most consistent and generally useful results.

Significant El Niño events have MEIs greater than 1; significant La Niña events have MEIs below -1.

Monthly ENSO Measures Since 1950

Appendix 7.1 summarizes the three measures of ENSO discussed in this chapter (the SOI, sea surface temperature anomalies in Region 3.4, and the (MEI).

El Niño and La Niña Years

The following lists of warm and cold events are based on Kiladis and Diaz's (1989) Climate Diagnostics Center NOAA CIRES for the years from 1875 to 1950 and on the combination of SOI, sea surface temperatures, and MEI from 1950 to 2000.

El Niño (Warm Event) Years: 1877–1878, 1880–1881, 1884–1885, 1891–1892, 1896–1897, 1899–1900, 1902–1903, 1904–1905, 1911–1912, 1913–1914, 1918–1919, 1923–1924, 1925–1926, 1930–1931, 1932–1933, 1939–1940, 1951–1952, 1953–1954, 1957–1958, 1958–1959, 1963–1964, 1965–1966, 1968–1969, 1969–1970, 1972–1973, 1976–1977, 1977–1978, 1979–1980, 1982–1983, 1986–1987, 1987–1988, 1991–1992, 1992–1993, 1994–1995, 1997–1998

La Niña (Cold Event) Years: 1886–1887, 1889–1890, 1892–1893, 1903–1904, 1906–1907, 1908–1909, 1916–1917, 1920–1921, 1924–1925, 1928–1929, 1931–1932, 1938–1939, 1942–1943, 1949–1950, 1950–1951, 1954–1955, 1955–1956, 1964–1965, 1970–1971, 1971–1972, 1973–1974, 1975–1976, 1983–1984, 1984–1985, 1988–1989, 1995–1996, 1998–1999, 1999–2000, 2000–2001.

As is often the case with El Niño and La Niña research, the years identified are not identical to the years indicated from 1877 to 1950 from Kiladis and Diaz (1989), as the sources may emphasize different measures. The best agreement, of course, occurs with the stronger events and the most disagreement with the weaker events in which the measures are more likely to disagree. There is perfect agreement for the years from 1950 to the present.

The Trans-Niño Index

Decadal scale changes in the thermal structures and circulations have been observed in both oceans (discussed further in chap. 10). The change in the Pacific, called the Pacific Decadal Oscillation (PDO), may have a key role in the frequency of El Niño and La Niña events. Similar changes in the Atlantic Ocean may influence the preferred phase of other circulation features in the atmosphere (e.g., the North Atlantic Oscillation).

In one attempt to measure some of the long-term changes in the Pacific, the Climate Analysis section of the National Center for Atmospheric Research (NCAR) developed the Trans-Nino Index (TNI; *Journal of Climate*, 2001). The TNI measures the difference in sea surface changes from the central to the eastern tropical Pacific. The TNI changed character abruptly after 1976–1977 (when the cold phase of the PDO gave way to the warm phase).

TRACKING ENSO

Monthly (or even weekly or daily) ENSO measures can be tracked on the World Wide Web. The following Web sites update these parameters regularly (URLs are subject to change).

SOI

The National Center on Environmental Prediction's Climate Prediction Center provides indices including SOI and sea surface temperature anomalies. Most of the indices are updated monthly (updated usually during the second week of the month) and include a historical database of SOI.

http://www.cpc.ncep.noaa.gov/data/indices
http://www.cpc.noaa.gov/data/indices/tahiti

Australian State of Queensland's Department of Natural Resources and the Australian Bureau of Meteorology provide daily (Monday through Friday) SOI updates along with running means for 30 and 90 days. The site also maintains a historical database of SOI back to 1900.

http://www.dnr.qld.gov.au/longpdk/latest/lattable.htm

Sea Surface Temperature Anomalies

Twice-weekly global sea surface temperature anomalies from NOAA's NESDIS based on satellite sensors. Historical maps are available for comparison.

http://psbsgi1.nesdis.noaa.gov:8080/PSB/EPS/SST/climo.html

U.S. Navy's daily sea surface temperature anomaly. This is a short-term forecast (NOWcast).

http://www.fnoc.navy.mil/PUBLIC/

Multi-Variate ENSO Index (MEI)

Climate Diagnostic's Center MEI is explained and a monthly update provided during ENSO events with the latest 2-month MEI compared to the strongest prior events.

http://www.cdc.noaa.gov/~kew/MEI/mei.html#outlook

Monthly ENSO Advisories

Climate Prediction Center provides monthly advisories on the status of ENSO events. Advisories often refer to, and interpret the latest monthly indices.

http://www.cpc.ncep.noaa.gov/products/analysis_monitoring/enso_advisory/

Appendix 7.1
Monthly ENSO Measures Since 1950

YEAR	Month	Southern Oscillation Index (SOI)	Sea Surface Temperature Anomaly (°C) Region 3.4	Multi-Variate ENSO Index (MEI)
1950	January	0.5	-1.33	-1.1
1950	February	2.1	-1.69	-1.2
1950	March	1.9	-0.66	-1.1
1950	April	1.2	-0.83	-1.2
1950	May	1.6	-1.23	-1.5
1950	June	2.0	-0.61	-1.4
1950	July	2.0	-0.37	-1.4
1950	August	1.1	-0.53	-1.2
1950	September	0.7	-0.84	-0.8
1950	October	1.6	-0.42	-0.5
1950	November	1.0	-0.93	-0.8
1950	December	2.7	-0.85	-1.2
1951	January	1.7	-0.87	-1.1
1951	February	0.6	+0.01	-1.2
1951	March	-0.8	-0.64	-0.8
1951	April	-0.6	+0.16	-0.4
1951	May	-1.0	+0.01	+0.1
1951	June	-0.3	+0.13	+0.6
1951	July	-1.4	+0.52	+0.6
1951	August	-0.7	+0.80	+0.9
1951	September	-1.3	+0.62	+0.9
1951	October	-1.4	+0.95	+0.8
1951	November	-1.0	+0.96	+0.7
1951	December	-1.0	+0.80	+0.6
1952	January	-1.2	+0.54	+0.3
1952	February	-1.1	+0.42	+0.2
1952	March	0.0	-0.19	+0.2
1952	April	-0.5	+0.54	0
1952	May	0.6	+0.05	-0.5
1952	June	0.5	-0.34	-0.4
1952	July	0.4	-0.14	-0.4
1952	August	0.3	+0.19	-0.2
1952	September	-0.3	+0.09	+0.1
1952	October	0.2	+0.25	+0.3
1952	November	-0.2	+0.01	0
1952	December	-1.6	-0.40	-0.2
1953	January	0.2	+0.68	0
1953	February	-1.0	+0.47	0.2
1953	March	-0.8	+0.33	0.3
1953	April	-0.1	+0.82	0.5
1953	May	-2.2	+0.37	0.9
1953	June	-0.3	+0.47	0.6
1953	July	-0.1	+0.69	0.4
1953	August	-1.9	+0.19	0.4
1953	September	-1.5	+1.06	0.3
1953	October	-0.2	+0.46	0.3
1953	November	-0.4	+0.59	0.1
1953	December	-0.7	+0.04	0.2

1954	January	0.6	+0.74	0
1954	February	-0.7	+0.33	0
1954	March	-0.3	+0.28	-0.3
1954	April	0.4	-0.19	-1.1
1954	May	0.3	-0.07	-1.5
1954	June	-0.3	-0.34	-1.5
1954	July	0.3	-0.64	-1.4
1954	August	0.8	-0.99	-1.3
1954	September	0.2	-0.84	-1.3
1954	October	0.1	-0.65	-1.2
1954	November	0.1	-0.71	-1.1
1954	December	1.5	-0.86	-0.9
1955	January	-0.7	-1.22	-0.7
1955	February	1.8	-0.34	-0.9
1955	March	0.1	-0.77	-1.4
1955	April	-0.5	-0.57	-1.6
1955	May	0.9	-0.87	-1.9
1955	June	1.1	-0.69	-2.0
1955	July	1.7	-0.57	-1.9
1955	August	1.2	-0.72	-1.9
1955	September	1.5	-1.39	-1.8
1955	October	1.5	-1.62	-1.9
1955	November	1.3	-2.02	-1.9
1955	December	1.0	-1.34	-1.7
1956	January	1.4	-1.02	-1.4
1956	February	1.5	-0.43	-1.4
1956	March	0.9	-0.71	-1.3
1956	April	0.7	-0.66	-1.2
1956	May	1.3	+0.11	-1.4
1956	June	0.8	-0.25	-1.3
1956	July	1.1	-0.64	-1.1
1956	August	0.9	-0.66	-1.2
1956	September	0.0	-0.61	-1.4
1956	October	1.9	-0.16	-1.3
1956	November	0.1	-0.73	-1.1
1956	December	1.0	-0.19	-1.0
1957	January	0.6	-0.18	-0.7
1957	February	-0.5	+0.27	-0.2
1957	March	-0.4	+0.29	0.2
1957	April	0.0	+0.59	0.5
1957	May	-1.0	+0.60	0.8
1957	June	-0.2	+0.50	0.9
1957	July	0.1	+0.86	1.0
1957	August	-1.0	+1.33	1.1
1957	September	-1.1	+0.60	1.1
1957	October	-0.2	+0.94	1.1
1957	November	-1.2	+1.57	1.1
1957	December	-0.5	+1.69	1.3
1958	January	-2.3	+2.07	1.4
1958	February	-1.0	+1.63	1.3
1958	March	-0.3	+1.23	1.0
1958	April	0.1	+0.64	0.8
1958	May	-0.9	+0.59	0.8

1958	June	-0.2	+0.70	0.8
1958	July	0.3	+0.37	0.6
1958	August	0.6	+0.58	0.2
1958	September	-0.4	-0.26	0.1
1958	October	-0.2	+0.25	0.3
1958	November	-0.6	+0.36	0.6
1958	December	-0.9	+0.63	0.6
1959	January	-1.2	+0.72	0.7
1959	February	-2.0	+0.68	0.6
1959	March	0.9	+0.16	0.3
1959	April	0.2	+0.39	0.1
1959	May	0.3	+0.39	0
1959	June	-0.6	-0.04	-0.1
1959	July	-0.5	-0.34	-0.1
1959	August	-0.6	-0.31	0.1
1959	September	0.0	-0.35	-0.1
1959	October	0.3	+0.07	-0.1
1959	November	1.0	+0.04	-0.2
1959	December	0.8	0.00	-0.3
1960	January	0.0	+0.14	-0.3
1960	February	-0.3	-0.27	-0.2
1960	March	0.6	+0.09	-0.1
1960	April	0.6	+0.26	-0.2
1960	May	0.3	+0.12	-0.3
1960	June	-0.3	-0.05	-0.3
1960	July	0.4	+0.11	-0.3
1960	August	0.5	+0.26	-0.4
1960	September	0.7	+0.23	-0.5
1960	October	-0.1	-0.17	-0.4
1960	November	0.5	-0.24	-0.4
1960	December	0.8	-0.08	-0.3
1961	January	-0.4	-0.12	-0.2
1961	February	0.7	0.00	-0.1
1961	March	-2.7	-0.05	0
1961	April	0.7	+0.18	-0.1
1961	May	0.1	+0.31	-0.2
1961	June	-0.3	+0.65	-0.2
1961	July	0.1	-0.05	-0.3
1961	August	-0.2	-0.19	-0.4
1961	September	0.1	-0.47	-0.5
1961	October	-0.7	-0.42	-0.5
1961	November	0.6	-0.20	-0.5
1961	December	1.6	-0.11	-0.9
1962	January	2.2	-0.17	-1.1
1962	February	-0.7	-0.25	-1.0
1962	March	-0.4	-0.21	-0.9
1962	April	0.0	-0.30	-1.0
1962	May	1.0	-0.29	-0.9
1962	June	0.4	0.00	-0.8
1962	July	-0.1	+0.09	-0.7
1962	August	0.3	+0.17	-0.5
1962	September	0.5	-0.30	-0.6
1962	October	0.9	-0.28	-0.7

1962	November	0.3	-0.39	-0.6
1962	December	0.0	-0.57	-0.6
1963	January	1.1	-0.47	-0.9
1963	February	0.4	-0.43	-0.8
1963	March	0.7	+0.07	-0.8
1963	April	0.6	+0.25	-0.6
1963	May	0.1	+0.03	-0.3
1963	June	-1.0	+0.20	0.2
1963	July	-0.3	+1.05	0.5
1963	August	-0.5	+0.98	0.7
1963	September	-0.7	+1.02	0.8
1963	October	-1.8	+1.14	0.9
1963	November	-1.0	+1.10	0.8
1963	December	-1.6	+1.19	0.8
1964	January	-0.5	+1.08	0.7
1964	February	-0.3	+0.53	0.1
1964	March	0.7	-0.12	-0.5
1964	April	1.0	-0.51	-1.0
1964	May	-0.1	-0.71	-1.2
1964	June	0.4	-0.82	-1.3
1964	July	0.4	-0.43	-1.5
1964	August	1.3	-0.64	-1.4
1964	September	1.4	-0.88	-1.3
1964	October	1.3	-0.62	-1.2
1964	November	0.0	-1.03	-1.1
1964	December	-0.5	-1.01	-0.7
1965	January	-0.6	-0.52	-0.4
1965	February	0.1	-0.20	-0.3
1965	March	0.2	+0.08	-0.1
1965	April	-0.8	+0.18	0.4
1965	May	-0.1	+0.57	0.7
1965	June	-1.0	+0.87	1.2
1965	July	-2.0	+1.27	1.5
1965	August	-1.2	+1.41	1.5
1965	September	-1.5	+1.54	1.3
1965	October	-1.2	+1.71	1.3
1965	November	-1.8	+1.78	1.3
1965	December	0.0	+1.92	1.3
1966	January	-1.7	+1.48	1.3
1966	February	-0.7	+1.06	0.9
1966	March	-1.7	+0.96	0.5
1966	April	-0.5	+0.81	0.2
1966	May	-0.7	+0.13	-0.2
1966	June	0.0	+0.62	-0.2
1966	July	-0.1	+0.49	0
1966	August	0.3	+0.15	0
1966	September	-0.3	+0.11	-0.1
1966	October	-0.4	+0.06	0
1966	November	-0.1	-0.07	-0.1
1966	December	-0.6	-0.22	-0.3
1967	January	1.9	-0.19	-0.7
1967	February	1.6	-0.20	-1.0
1967	March	0.8	-0.50	-1.1

1967	April	-0.3	-0.71	-0.7
1967	May	-0.3	-0.03	-0.4
1967	June	0.3	+0.33	-0.5
1967	July	0.0	-0.03	-0.6
1967	August	0.5	-0.16	-0.6
1967	September	0.6	-0.36	-0.7
1967	October	-0.2	-0.36	-0.6
1967	November	-0.6	-0.17	-0.4
1967	December	-0.8	-0.28	-0.5
1968	January	0.4	-0.46	-0.8
1968	February	1.1	-0.85	-0.8
1968	March	-0.5	-0.53	-0.9
1968	April	-0.2	-0.30	-1.0
1968	May	1.1	-0.33	-0.9
1968	June	0.9	+0.20	-0.7
1968	July	0.6	+0.51	-0.3
1968	August	-0.1	+0.50	0
1968	September	-0.3	+0.29	0.3
1968	October	-0.3	+0.47	0.5
1968	November	-0.5	+1.00	0.5
1968	December	0.0	+0.88	0.5
1969	January	-2.0	+1.35	0.7
1969	February	-1.1	+1.19	0.6
1969	March	-0.1	+0.91	0.5
1969	April	-0.6	+0.60	0.6
1969	May	-0.6	+1.04	0.8
1969	June	-0.2	+0.69	0.6
1969	July	-0.7	+0.32	0.3
1969	August	-0.6	+0.69	0.3
1969	September	-1.2	+0.87	0.4
1969	October	-1.3	+0.99	0.6
1969	November	-0.2	+0.89	0.5
1969	December	0.3	+1.23	0.4
1970	January	-1.4	+1.05	0.4
1970	February	-1.6	+0.49	0.3
1970	March	0.0	+0.34	0.1
1970	April	-0.4	+0.56	-0.1
1970	May	0.1	+0.16	-0.4
1970	June	0.7	-0.31	-0.9
1970	July	-0.6	-0.75	-1.1
1970	August	0.2	-0.81	-1.1
1970	September	1.3	-0.82	-1.2
1970	October	0.9	-1.04	-1.1
1970	November	1.7	-1.29	-1.2
1970	December	2.1	-1.76	-1.2
1971	January	0.3	-1.33	-1.4
1971	February	1.9	-1.34	-1.6
1971	March	2.1	-0.97	-1.8
1971	April	1.7	-0.94	-1.6
1971	May	0.7	-0.36	-1.5
1971	June	0.1	-0.47	-1.3
1971	July	0.1	-0.32	-1.1
1971	August	1.3	-0.39	-1.1

1971	September	1.6	-0.52	-1.2
1971	October	1.7	-0.53	-1.1
1971	November	0.5	-0.78	-1.2
1971	December	0.0	-0.78	-0.8
1972	January	0.4	-0.50	-0.5
1972	February	0.8	-0.13	-0.3
1972	March	0.1	0.00	-0.2
1972	April	-0.4	+0.52	0.2
1972	May	-2.1	+0.83	0.8
1972	June	-1.1	+0.94	1.5
1972	July	-1.9	+1.21	1.8
1972	August	-1.0	+1.50	1.7
1972	September	-1.6	+1.46	1.6
1972	October	-1.2	+1.93	1.7
1972	November	-0.5	+2.27	1.7
1972	December	-1.6	+2.22	1.8
1973	January	-0.4	+1.91	1.7
1973	February	-2.0	+1.37	1.3
1973	March	0.2	+0.75	0.7
1973	April	-0.2	-0.05	0.2
1973	May	0.2	-0.27	-0.4
1973	June	0.8	-0.70	-0.9
1973	July	0.5	0.92	-1.2
1973	August	1.1	-0.98	-1.5
1973	September	1.4	-0.96	-1.7
1973	October	0.6	-1.21	-1.6
1973	November	2.0	-1.59	-1.6
1973	December	2.0	-1.81	-1.6
1974	January	2.7	-1.76	-1.9
1974	February	2.0	-1.41	-1.8
1974	March	2.2	-1.03	-1.8
1974	April	0.8	-0.68	-1.6
1974	May	0.9	-0.71	-1.3
1974	June	0.1	-0.22	-0.8
1974	July	1.2	-0.25	-0.7
1974	August	0.5	-0.33	-0.8
1974	September	1.3	-0.21	-0.7
1974	October	0.8	-0.58	-0.8
1974	November	-0.3	-0.77	-1.1
1974	December	0.0	-0.61	-1.0
1975	January	-0.8	-0.26	-0.7
1975	February	0.6	-0.29	-0.5
1975	March	1.2	-0.48	-0.7
1975	April	1.1	-0.43	-0.9
1975	May	0.5	-0.65	-0.9
1975	June	1.1	-1.00	-1.0
1975	July	2.1	-0.83	-1.6
1975	August	1.9	-1.09	-1.8
1975	September	2.4	-1.09	-1.9
1975	October	1.7	-1.38	-1.9
1975	November	1.3	-1.12	-1.8
1975	December	2.3	-1.57	-1.7
1976	January	1.5	-1.68	-1.5

1976	February	1.6	-1.00	-1.3
1976	March	1.3	-0.59	-1.2
1976	April	0.1	-0.59	-0.8
1976	May	0.2	-0.31	-0.1
1976	June	-0.1	-0.01	0.4
1976	July	-1.2	+0.25	0.7
1976	August	-1.3	+0.43	0.9
1976	September	-1.4	+0.80	1.0
1976	October	0.2	+1.13	0.7
1976	November	0.7	+1.12	0.5
1976	December	-0.6	+0.82	0.5
1977	January	-0.7	+1.04	0.4
1977	February	1.1	+0.41	0.3
1977	March	-1.3	+0.46	0.4
1977	April	-0.8	0.00	0.5
1977	May	-0.9	+0.50	0.4
1977	June	-1.5	+0.46	0.6
1977	July	-1.5	+0.42	0.8
1977	August	-1.3	+0.21	0.7
1977	September	-1.0	+0.56	0.9
1977	October	-1.4	+0.85	1.0
1977	November	-1.6	+0.93	0.9
1977	December	-1.4	+0.77	0.8
1978	January	-0.4	+1.00	0.8
1978	February	-3.5	+0.32	0.9
1978	March	-0.8	+0.28	0.5
1978	April	-0.6	-0.36	-0.1
1978	May	1.3	-0.14	-0.5
1978	June	0.3	-0.29	-0.4
1978	July	0.4	-0.28	-0.3
1978	August	0.0	-0.51	-0.3
1978	September	0.0	-0.19	-0.2
1978	October	-0.7	-0.04	0.1
1978	November	-0.1	+0.05	0.3
1978	December	-0.3	+0.14	0.5
1979	January	-0.7	+0.16	0.5
1979	February	0.8	+0.08	0.2
1979	March	-0.5	+0.53	0.2
1979	April	-0.4	+0.31	0.4
1979	May	0.3	+0.29	0.5
1979	June	0.4	+0.36	0.4
1979	July	1.3	-0.12	0.5
1979	August	-0.6	+0.05	0.7
1979	September	0.1	+1.11	0.8
1979	October	-0.4	+0.42	0.7
1979	November	-0.6	+0.53	0.9
1979	December	-1.0	+0.64	0.8
1980	January	0.3	+0.79	0.6
1980	February	0.0	+0.56	0.6
1980	March	-1.2	+0.20	0.8
1980	April	-1.0	+0.28	0.9
1980	May	-0.3	+0.43	0.9
1980	June	-0.4	+0.72	0.9

1980	July	-0.2	+0.36	0.6
1980	August	0.0	-0.05	0.3
1980	September	-0.6	+0.20	0.2
1980	October	-0.3	+0.05	0.2
1980	November	-0.5	+0.25	0.1
1980	December	-0.3	+0.52	-0.2
1981	January	0.2	-0.23	-0.3
1981	February	-0.6	-0.45	0.1
1981	March	-2.1	-0.16	0.5
1981	April	-0.4	-0.30	0.3
1981	May	0.7	-0.07	0
1981	June	1.0	+0.01	0
1981	July	0.8	-0.31	-0.1
1981	August	0.4	-0.63	0
1981	September	0.4	+0.17	0.1
1981	October	-0.7	+0.35	0.1
1981	November	0.1	+0.06	-0.1
1981	December	0.5	+0.19	-0.2
1982	January	1.3	+0.36	-0.3
1982	February	-0.1	+0.06	-0.1
1982	March	0.1	+0.09	0
1982	April	-0.2	+0.39	0.2
1982	May	-0.7	+0.99	0.6
1982	June	-1.6	+1.33	1.3
1982	July	-1.9	+1.14	1.7
1982	August	-2.5	+1.33	1.8
1982	September	-2.0	+1,78	1.9
1982	October	-2.2	+2.27	2.2
1982	November	-3.2	+2.46	2.5
1982	December	-2.8	+2.78	2.7
1983	January	-4.2	+3.01	2.9
1983	February	-4.6	+2.62	3.1
1983	March	-3.4	+1.98	3.1
1983	April	-1.3	+1.35	2.8
1983	May	0.5	+1.40	2.4
1983	June	-0.3	+0.90	2.0
1983	July	-0.8	+0.12	1.5
1983	August	-0.2	-0.09	0.8
1983	September	1.0	+0.05	0.3
1983	October	0.3	-0.61	0
1983	November	-0.2	-0.72	-0.1
1983	December	-0.1	-0.83	-0.2
1984	January	0.1	-0.73	-0.4
1984	February	0.6	-0.28	-0.1
1984	March	-0.9	-0.25	0.4
1984	April	0.2	-0.18	0.3
1984	May	0.0	-0.14	0
1984	June	-0.8	-0.53	-0.2
1984	July	0.0	-0.25	-0.2
1984	August	0.0	-0.35	-0.1
1984	September	0.1	+0.01	0
1984	October	-0.6	-0.55	-0.2
1984	November	0.2	-0.95	-0.4

1984	December	-0.4	-1.46	-0.6
1985	January	-0.5	-0.81	-0.6
1985	February	1.0	-1.06	-0.6
1985	March	0.2	-0.86	-0.6
1985	April	1.0	-0.69	-0.6
1985	May	0.2	-0.41	-0.4
1985	June	-0.9	-0.47	-0.2
1985	July	-0.3	-0.25	-0.3
1985	August	0.7	-0.18	-0.5
1985	September	0.0	-0.19	-0.3
1985	October	-0.7	-0.32	-0.1
1985	November	-0.3	-0.20	-0.2
1985	December	0.1	-0.22	-0.3
1986	January	0.9	-0.60	-0.3
1986	February	-1.6	-0.73	-0.1
1986	March	0.0	-0.46	0
1986	April	0.1	-0.12	0.1
1986	May	-0.5	-0.03	0.3
1986	June	-0.7	+0.25	0.4
1986	July	0.1	+0.39	0.5
1986	August	-1.0	+0.55	0.9
1986	September	-0.6	+0.98	1.1
1986	October	0.5	+1.10	0.9
1986	November	-1.5	+1.31	1.0
1986	December	-1.8	+1.29	1.2
1987	January	-0.9	+1.53	1.2
1987	February	-1.9	+1.41	1.4
1987	March	-2.0	+1.39	1.7
1987	April	-1.9	+1.19	2.0
1987	May	-1.7	+1.21	2.0
1987	June	-1.7	+1.61	1.9
1987	July	-1.7	+1.83	1.9
1987	August	-1.5	+1.96	1.9
1987	September	-1.2	+1.98	1.8
1987	October	-0.7	+1.63	1.4
1987	November	-0.1	+1.59	1.3
1987	December	-0.7	+1.18	1.2
1988	January	-0.2	+0.96	0.9
1988	February	-0.9	+0.57	0.6
1988	March	0.1	+0.26	0.5
1988	April	-0.1	-0.26	0.2
1988	May	0.8	-1.03	-0.3
1988	June	-0.2	-1.31	-0.9
1988	July	1.1	-1.38	-1.3
1988	August	1.4	-1.42	-1.4
1988	September	2.1	-1.00	-1.5
1988	October	1.4	-1.79	-1.4
1988	November	1.9	-2.09	-1.4
1988	December	1.3	-2.03	-1.2
1989	January	1.7	-1.85	-1.1
1989	February	1.1	-1.30	-1.1
1989	March	0.6	-1.22	-0.8
1989	April	1.6	-0.89	-0.6

1989	May	1.2	-0.44	-0.4
1989	June	0.5	-0.40	-0.4
1989	July	0.8	-0.24	-0.5
1989	August	-0.8	-0.27	-0.5
1989	September	0.6	-0.14	-0.4
1989	October	0.6	-0.20	-0.3
1989	November	-0.4	-0.14	-0.2
1989	December	-0.7	-0.04	0.1
1990	January	-0.2	+0.14	0.2
1990	February	-2.4	+0.26	0.4
1990	March	-1.2	+0.36	0.7
1990	April	0.0	+0.45	0.6
1990	May	1.1	+0.55	0.4
1990	June	0.0	+0.21	0.5
1990	July	0.5	+0.34	0.3
1990	August	-0.6	+0.41	0.1
1990	September	-0.8	+0.37	0.3
1990	October	0.1	+0.50	0.3
1990	November	-0.7	+0.34	0.3
1990	December	-0.5	+0.47	0.4
1991	January	0.6	+0.62	0.3
1991	February	-0.1	+0.30	0.3
1991	March	-1.4	+0.18	0.3
1991	April	-1.0	+0.39	0.3
1991	May	-1.5	+0.84	0.5
1991	June	-0.5	+0.98	0.9
1991	July	-0.2	+0.96	1.0
1991	August	-0.9	+0.72	0.8
1991	September	-1.8	+0.64	0.8
1991	October	-1.5	+1.24	1.1
1991	November	-0.8	+1.51	1.2
1991	December	-2.3	+1.93	1.5
1992	January	-3.4	+2.00	1.8
1992	February	-1.4	+1.99	1.9
1992	March	-3.0	+1.73	2.1
1992	April	-1.4	+1.58	2.2
1992	May	0.0	+1.46	2.0
1992	June	-1.2	+0.67	1.4
1992	July	-0.8	+0.58	0.8
1992	August	0.0	+0.07	0.6
1992	September	0.0	+0.15	0.6
1992	October	-1.9	-0.05	0.6
1992	November	-0.9	+0.14	0.6
1992	December	-0.9	+0.34	0.6
1993	January	-1.2	+0.33	0.8
1993	February	-1.3	+0.32	0.9
1993	March	-1.1	+0.54	1.1
1993	April	-1.6	+1.01	1.7
1993	May	-0.6	+1.31	1.7
1993	June	-1.4	+0.86	1.4
1993	July	-1.1	+0.60	1.1
1993	August	-1.5	+0.23	1.0
1993	September	-0.8	+0.52	1.0

1993	October	-1.5	+0.50	0.9
1993	November	-0.8	+0.48	0.7
1993	December	-2.3	+0.34	0.5
1994	January	-0.3	+0.18	0.3
1994	February	-0.1	-0.08	0.2
1994	March	-1.4	+0.09	0.3
1994	April	-1.8	+0.23	0.5
1994	May	-1.0	+0.49	0.6
1994	June	-0.9	+0.51	0.7
1994	July	-1.8	+0.35	0.7
1994	August	-1.8	+0.67	0.6
1994	September	-1.8	+0.62	1.0
1994	October	-1.6	+0.93	1.3
1994	November	-0.7	+1.44	1.2
1994	December	-1.6	+1.43	1.2
1995	January	-0.6	+1.15	1.0
1995	February	-0.5	+0.83	0.8
1995	March	0.2	+0.45	0.4
1995	April	-1.1	+0.33	0.4
1995	May	-0.7	+0.14	0.4
1995	June	-0.2	+0.12	0.4
1995	July	0.3	+0.06	0.2
1995	August	-0.1	-0.26	-0.2
1995	September	0.3	-0.47	-0.4
1995	October	-0.3	-0.78	-0.4
1995	November	0.0	-0.77	-0.5
1995	December	-0.8	-0.82	-0.6
1996	January	1.0	-0.66	-0.6
1996	February	-0.1	-0.81	-0.5
1996	March	0.7	-0.43	-0.4
1996	April	0.6	-0.19	-0.4
1996	May	0.1	-0.20	-0.1
1996	June	1.0	-0.06	-0.1
1996	July	0.6	+0.13	-0.2
1996	August	0.4	-0.08	-0.3
1996	September	0.6	-0.07	-0.3
1996	October	0.4	-0.22	-0.2
1996	November	-0.2	-0.18	-0.2
1996	December	0.8	-0.42	-0.4
1997	January	0.5	-0.40	-0.5
1997	February	1.6	-0.25	-0.3
1997	March	-1.1	+0.03	0.1
1997	April	-0.9	+0.51	0.7
1997	May	-1.8	+1.10	1.7
1997	June	-2.0	+1.56	2.5
1997	July	-1.0	+1.96	2.7
1997	August	-2.1	+2.20	2.8
1997	September	-1.6	+2.53	2.5
1997	October	-1.9	+2.79	2.3
1997	November	-1.4	+2.92	2.3
1997	December	-1.3	+2.85	2.3
1998	January	-3.3	+2.75	2.6
1998	February	-2.7	+2.25	2.8

1998	March	-3.5	+1.59	2.7
1998	April	-1.9	+0.96	2.3
1998	May	0.1	+0.93	1.6
1998	June	0.7	-0.69	0.7
1998	July	1.3	-1.02	0.1
1998	August	1.0	-1.15	-0.4
1998	September	1.2	-0.79	-0.7
1998	October	1.0	-1.11	-0.9
1998	November	1.1	-1.18	-1.0
1998	December	1.4	-1.61	-1.0
1999	January	2.0	-1.45	-1.0
1999	February	0.8	-1.20	-0.9
1999	March	0.9	-0.83	-0.8
1999	April	1.4	-0.73	-0.7
1999	May	0.1	-0.65	-0.5
1999	June	-0.1	-0.88	-0.4
1999	July	0.5	-0.67	-0.6
1999	August	0.1	-1.06	-0.8
1999	September	-0.1	-0.86	-0.9
1999	October	0.9	-0.88	-1.0
1999	November	1.1	-1.31	-1.1
1999	December	1.5	-1.57	-1.2
2000	January	0.6	-1.79	-1.1
2000	February	1.6	-1.46	-1.0
2000	March	1	-1.03	-0.7
2000	April	1.2	-0.56	-0.2
2000	May	0.2	-0.50	-0.2
2000	June	-0.6	-0.44	-0.3
2000	July	-0.4	-0.30	-0.3
2000	August	0.4	-0.20	-0.2
2000	September	1.0	-0.36	-0.3
2000	October	1.0	-0.55	-0.6
2000	November	2.0	-0.65	-0.8
2000	December	0.7	-0.88	-0.7

Data from NOAA. MEI is approximated from bi-monthly average data provided.

Source: Sea surface temperature anomalies and SOI from NOAA at http://www.cpc.ncep. noaa.gov/~kew/MEI/mei.html#outlook. Courtesy of National Oceanic & Atmospheric Administration (NOAA).

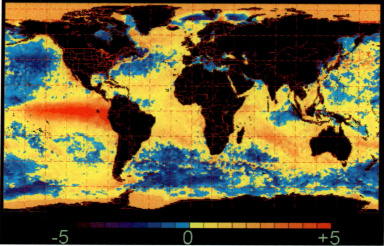

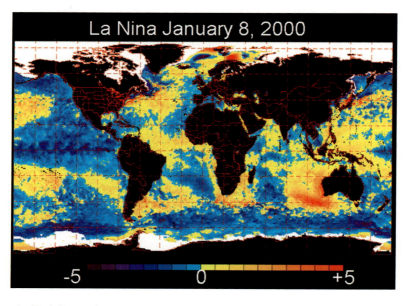

Plate 1. *(Top)* Sea surface temperature anomalies on January 24, 1998 during the Great El Niño of 1997/1998. Warmer than normal ocean water is shown in yellow and red, and colder than normal ocean water is shown in shades of blue. Note the warm plume of water extending west off of the west coast of South America along the equator characteristic of the warm El Niño phase. *(Bottom)* Sea surface temperature anomalies on January 8, 2000 during the subsequent La Niña. Note the colder than normal ocean water, shown in shades of blue, extending west off the west coast of South America along the equator character-istic of the cold La Niña phase. *Courtesy of NOAA National Environmental Satellite, Data, and Information Service (NESDIS).*

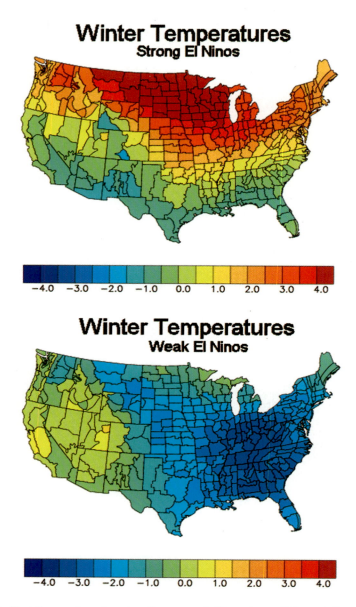

Plate 2. *(Top)* Temperature departure from normal (December to February) for strong El Niño years. Note the warmth centered over the north central states, with cool weather restricted to the cloudy and wet southern states. *(Bottom)* Temperature departure from normal (December to February) for weak El Niño years. A very different temperature pattern results in weak El Niño. With less warming from the Pacific and a jet stream that dips into the central and eastern states, most of the central and eastern states are colder than normal. *Courtesy of NOAA-CIRES Climate Diagnostics Center.*

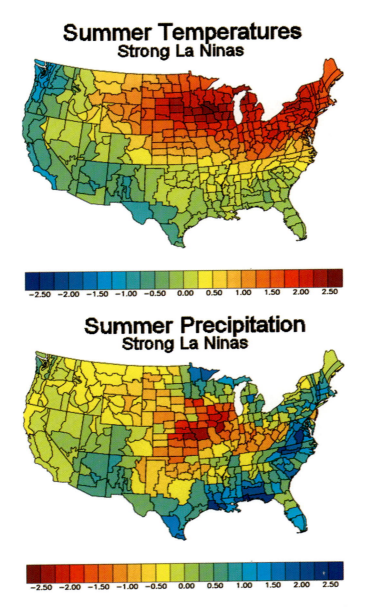

Plate 3. *(Top)* The Composite temperature anomalies for July to August for strong/strengthening La Niña years (1949, 1955, 1964, 1973, 1975, 1988, 1999). *(Bottom)* The Composite precipitation anomalies for July to August for strong/ strengthening La Niña years. *Courtesy of NOAA-CIRES Climate Diagnostic Center.*

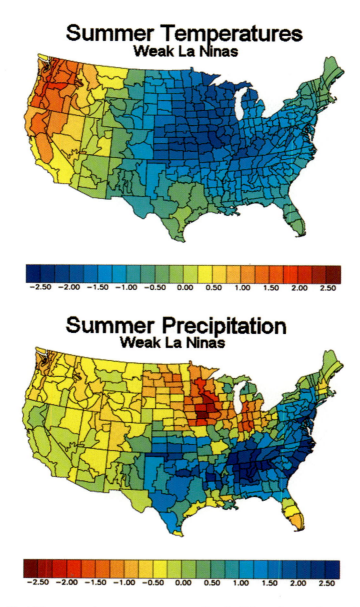

Plate 4. *(Top)* The composite temperature anomalies for July to August for weak/ weakening La Niña years (1950, 1956, 1967, 1971, 1974, 1985, 1989, 1996). *(Bottom)* The composite precipitation anomalies for July to August for weak/weakening La Niña years. *Courtesy of NOAA-CIRES-Climate Diagnostic Center.*

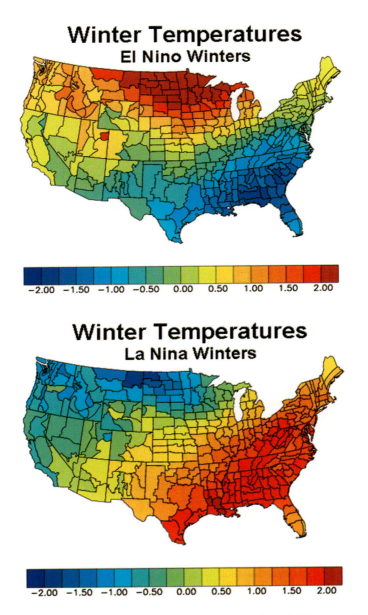

Plate 5. *(Top)* Composite winter temperatures (December through February) for all El Niño winters since 1950. Note the tendency for warmer than normal conditions in the north central and cooler in the Southeast. *(Bottom)* Composite winter temperatures (December through February) for La Nada winters that followed El Niños (1952/53, 1959/60, 1978/79, 1980/81, 1993/94). Note the coolness in the Southeast and warmth in the Northwest. *Courtesy of NOAA-CIRES-Climate Diagnostic Center.*

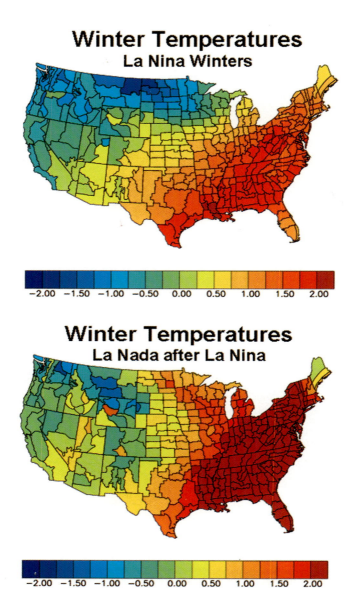

Plate 6. *(Top)* Composite winter temperatures (December through February) for all La Niña winters since 1950. Note the tendency for warmer than normal conditions in the Southeast and cooler in the Northwest and north central states. *(Bottom)* Composite winter temperatures (December through February) for La Nada winters that followed Na Niñas (1956/57, 1974/75, 1989/90, 1996/97). Note the warmth in the Southeast. *Courtesy of NOAA-CIRES-Climate Diagnostic Center.*

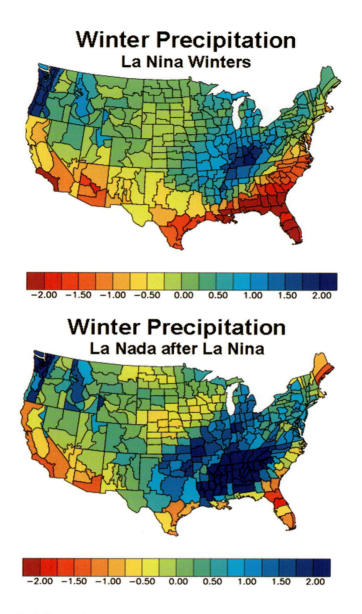

Plate 7. *(Top)* Composite winter precipitation (December through February) for all La Niña winters since 1950. Note the tendency for wetter than normal weather in the northwest and the Tennessee and Ohio valleys with dryness along the immediate Gulf Coast. *(Bottom)* Composite Winter (December through February) Precipitation in La Nada winters that followed La Niñas (1956/57, 1974/75, 1989/90, 1996/97). A very similar pattern to the composite La Niña is observed with a tendency for wet weather in the central Gulf States, Tennessee Valley and Pacific Northwest. *Courtesy of NOAA-CIRES-Climate Diagnostic Center.*

Sea Surface Temperature Change

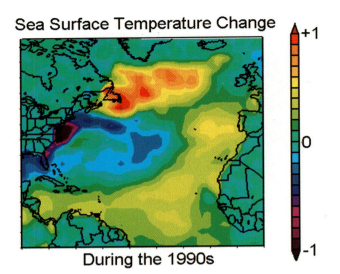

During the 1990s

Plate 8a. Changes in sea surface temperatures in the Atlantic from the early (1991 to 1995) to late (1996 to 2000) periods. Note the warming in the far north and near the equator, a pattern suggestive of changes in the thermohaline circulation in the Atlantic Basin back to a pattern seen frequently in the 1950s and 1960s. *Courtesy of NOAA-CIRES.*

Pacific Decadal Oscillation (PDO)

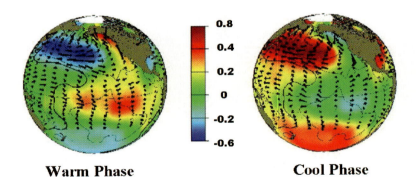

Warm Phase   **Cool Phase**

Plate 8b. The so-called "warm" phase of the PDO (left) has warm water in the central Tropical Pacific ringed by a horseshoe shaped cold water mass. In the "cold" phase (right), colder than normal tropical ocean water is ringed by water that is warmer than normal. *Courtesy of University of Washington.*

Chapter 8

Forecasting ENSO

Newly found knowledge of ENSO and its effects has made seasonal forecasts possible on a global scale. It is now believed that ENSO is itself predictable.

Two key reasons for the new predictability are improved observations and better statistical and numerical atmospheric and oceanic models. Also important is that, thanks to the research over the last half century, researchers now have a better understanding of what happens in the warm and cold event cycles and thus are better able to interpret the observations and model forecasts.

THE VALUE OF CLIMATE FORECASTS

The new predictability that this understanding of ENSO provides, will bring value to the world's economies with the greatest benefits to agriculture, water resource management, energy, public health, and safety. Countries most affected by ENSO, such as Australia, Peru, and Brazil, demonstrated in the 1980s the value of incorporating ENSO-related forecast information into agricultural and resource management decisions.

Dr. Tim Barnett of Scripps Institute of Oceanography, whose work on ENSO predictability dates back to the early 1980s (Barnett 1984) gave testimony to Congress on September 11, 1997 on how climate forecasts are being used. He told them: "Agricultural planners in Australia, Brazil and Peru used these forecasts to develop appropriate guidance products advising farmers to adjust fertilizers, planting schedules and/or crop types to avoid crop failure and maximize yields. They demonstrated that use of seasonal forecasts have been key in protecting and enhancing production and profit margins during periods of environmental stress."

Many other countries, including the United States, have followed suit. It is estimated that better climate predictions, if used appropriately, could reduce climate damage costs in the United States alone by 25 percent, or $2.7 billion annually in the agricultural sector alone. Water, energy, and transportation could also avoid or mitigate losses with proper use of accurate and timely forecasts.

In the autumn of 1997, Ants Leetma, the director of the NCEP, warned Californians of a long winter of powerful storms comparable to the winter of 1982–1983. He indicated that "The southern part of the state can expect rainfall on the order of 200% of normal." California took seriously his warnings that storms and flooding would be similar to the Great El Niño of 1982–1983, which produced about $2 billion in damages. The state of California spent $7.5 million to aid in preparedness and alerting the public. The southern parts of the state did in fact have a very stormy winter with about 230 percent of normal rainfall. Despite the continuous assault of storms from the Pacific, the $1.1 billion in losses were significantly less than in 1982–1983.

Ironically, industry and the public may lag local governments in accepting and acting on climate predictions. Research from the Climate Diagnostics Center has identified some possible barriers to seasonal forecasts being used for decision making. Among the reasons are the following:

1. Forecasts are not perceived "accurate" enough. Many believe that seasonal forecasts cannot be accurate because there are still errors in short-range forecasts.

2. Fluctuations of successive forecasts when it occurs ("waffling") lowers confidence.

3. Procedures for acquiring information and implementing decisions have not been clearly defined.

4. Desired information not provided (e.g., forecasts of how temperature and rainfall will likely compare to normal) or is not sufficient. Many instead want specific information such as number of 90F + days, number of heavy snow days, and so on.

5. History of previous forecasts (validation of accuracy) not available. Actual, written statistics carry weight.

Once informed and educated users "see the light," they can turn information to their advantage. Large institutions and businesses can save money. One example is a research project by meteorology students and faculty at Northern Illinois University that saved the university an estimated $500,000 in natural gas costs for the winter of 1997–1998. The group used climate profiles of the local region, historical natural gas usage, and ENSO forecast models to decide whether to lock in a price for energy early or ride out the market. Their research determined that warmer than normal conditions were expected for the season and the plant manager decided to ride the market and purchase the natural gas as needed rather than lock in a price. Prices fell through the season as warmer than normal weather materialized and reduced demand.

FORECASTING PROGRAMS

Tropical Ocean Global Atmosphere (TOGA) Program (1985–1995)

Over its 10-year lifetime, the Tropical Ocean Global Atmosphere (TOGA) program made major strides toward an understanding of the ENSO phenomenon. It demonstrated the feasibility of operational, multiseason climate prediction of ocean temperatures based on numerical models, and clarified the nature of the remote, planetary-scale atmospheric response to these anomalies. It was during this period that the greatest advances in the understanding of the ENSO phenomenon took place.

Part of TOGA's legacy lives on through NOAA's TAO Observing System and modern predictive capabilities of NOAA's Climate Prediction Center (CPC) and the International Research Institute for Climate Prediction (IRI) supported by both NOAA and international partner contributions.

More Accurate Observations

Today, thanks to technology, conditions in both the atmosphere and the oceans can be tracked in real time. The principal components of this ENSO Observing System are fixed buoys (moorings), volunteer ship temperature probes, and satellite observations.

An array of approximately seventy moorings (TAO/TRITON) measures ocean temperatures to a depth of 500 meters over the entire equatorial Pacific. This array is supported primarily by the United States (NOAA) and Japan (JAMSTEC) with contributions from France (IRD) and Taiwan (NTU). Oceanographic and meteorological data is then telemetered to shore in real time via the Argus satellite system. To supplement the buoys, volunteer ships measure ocean temperatures. The cost of these direct (buoy and ship) ocean observations is $4.9 million per year. As an important part of the monitoring effort, weather satellites measure ocean surface temperatures from space and also monitor surface and atmospheric winds, cloudiness, and even sea-level heights. With these tools, scientists can track the changes that precede the onset of a warm or cold event.

Because El Niño and La Niña involve slowly changing, large-scale circulations of water and air in the Tropical Pacific, they develop, evolve, and decay slowly. They average a year or more in length. Once a warm or cold event begins, it usually can be counted on to run its course over the following twelve months.

To attempt to provide advance indication of the start or end of an event and to determine how strong a developing event will be, scientists and governments have invested in the development of special ENSO computer models.

ENSO Models

There are a number of seasonal models used to predict the ENSO phenomenon. Computer models that are used to predict the atmospheric state (weather) over the short and medium-range (out to fourteen days) deal with the initial conditions and forecasts for the earth's surface and atmosphere. The seasonal forecast models require initial conditions and forecasts for both the atmosphere and the ocean. Many ENSO forecast techniques employ "coupled" global atmospheric and oceanic models. Statistical models have also demonstrated some skill in predicting future ENSO states using historical relationships. And there is also a set of models that combine these statistical and numerical model approaches.

Coupled Ocean and Atmospheric Models

Research into ENSO has demonstrated the importance that large bodies of unusually warm or cool water have to the atmospheric weather regimes and patterns. Although the oceanic state changes slowly relative to the air or ground, the changes are nonetheless important. The performance of models that assume a steady state for the ocean (the short-range forecast models) tend to degrade with time as the ocean changes.

Coupled ocean and atmospheric models predict changes in both the atmosphere and oceans. However, like the atmospheric models, the ocean models are not always right. In an attempt to establish the range of possibilities and determine the stability of a model forecast, the models are often run multiple times with slight perturbations to the initial state. This technique is called *ensembling* and the forecasts are called *ensemble forecasts*. The resulting solutions provide a "range" of possibilities. The more consistent the solutions, the higher the forecaster's confidence that the climate will evolve as the model predicts.

Zebiak and Cane LDEO Model

The first sea surface temperature forecasts were produced in 1987 (Zebiak and Cane, 1987) with a coupled atmospheric-ocean model (LDEO). Each month, historical wind data from 1964 to the latest month were used to "spin-up" the ocean model. In parallel, the atmospheric model was run with input from the ocean model. Then the two were run forward in time producing a forecast for the evolution of the couple atmospheric-ocean system. This model continues to be run today with improvements made to how the model is in initialized.

NCEP Coupled Model—Ensemble

This is a suite of global, coupled oceanic-atmospheric models run each month to provide seasonal forecasts for 10 months or more into the future. NCEP uses a state-of-the-art ocean data assimilation system to provide initial conditions for the ocean component of the coupled ocean atmosphere model. Analyses of sea surface temperatures are made weekly using a process known as optimum interpolation. In the interpolation technique, ocean temperature data is derived from buoys, ships, and satellite. The satellite-derived data is corrected for biases using methods developed by Reynolds and Smith (1994).

The data is then input into a coupled oceanic and atmospheric system, which then forecasts the oceanic and atmospheric conditions for the next 10 or more months.

Over the course of a month, the models are run 20 times with different weekly observational initial conditions (both oceanic and atmospheric conditions). The forecasts are subtracted from model-derived climatology (normal values) for each field to yield the anomalies (departures from normal). The mean anomalies of the forecast fields are then utilized for the actual climate forecasts. The greater the anomalies and the more consistent the results, the more significant the mean anomalies and, in general, the higher the confidence in the accuracy of the forecasts.

ECMWF Coupled Model

The European Center for Medium Range Wealth Forecasts (ECMWF) approved a program for seasonal prediction in 1994. In 1995, the group was assembled and seasonal prediction development began. In their approach, the initial state of the ocean is analyzed using fluxes of momentum (wind), heat (temperature), and fresh water (river outflows, ice or snowmelt, precipitation). In addition, thermal data from buoy and ship data is used, primarily from TOGA in the tropics and WOCE in the extratropical areas. NCEP's sea surface temperature analysis is also factored in.

The ECMWF coupled model consists of the ECMWF atmospheric model coupled to an ocean general circulation model (a version of the Hamburg Ocean Primitive Equation Model [HOPE] developed at the Max-Planck Institute for Meteorology in Hamburg). The atmosphere and ocean communicate with each other through a coupling interface. Three times a week, the coupled model is integrated forward for 6 months. This creates an ensemble or family of twelve to fifteen model forecasts each month. From these forecasts, forecast products, including sea surface temperature anomalies and the predicted atmospheric response, are generated.

Statistical Models

Climate Diagnostic Center's Linear Inverse Model

This is a statistical model that is used to predict sea surface temperature anomalies in the Tropical Pacific. It uses predictors in three geographical locations, the Tropical Pacific, the Tropical Pacific and Indian Oceans taken together and the global tropical oceans. The model looks at sea surface temperature patterns in the three regions and separates out the principal components of oscillation (called modes) that are apparent in the historical record. It has been shown that groups of normal modes can be combined to predict a measure field in a geographical area. This technique is used to predict sea surface temperature anomalies in NINO Region 3 of the Pacific.

The technique works best when the sea surface temperature anomalies are rapidly evolving. It performs poorly in situations where the ENSO events are prolonged (both warm and cold).

COLA Anomaly Coupled Prediction System

The Center for Ocean and Atmosphere Studies (COLA) developed an anomaly coupled prediction system, using sophisticated dynamical ocean and atmospheric models, that produce skillful sea surface temperature anomaly forecasts up to 18 months in advance.

The atmospheric forecasts are provided by COLA's global model, and the ocean forecasts by a Pacific Basin version of the Geophysical Fluid Dynamics Laboratory (GFDL) ocean model. Through experimentation, COLA developed a unique method for initializing the model using wind stresses at around 5,000 feet and for coupling atmosphere and ocean.

Verification of the model using reconstructed hindcasts for prior years showed the COLA model was competitive with any other coupled model used.

Ensemble Neural Network Model

This statistical ENSO prediction approach, developed at the University of British Columbia (Tangang, Hsieh, Tang, and Monahan, 1998) employs the historical patterns of Tropical Pacific sea-level pressure and sea surface temperatures. The data is smoothed (3-month running means) and then the data is analyzed into to a series of modes called Empirical Orthogonal Functions (EOFs). The principal modes are retained and normalized.

Every month, the EOF values for the current month and for three prior periods (3, 6, and 9 months before this month). This yields a series of time series that are input into a neural network. The process is repeated 20 times, each with a different weight initialization. The final forecast is the average of the outputs from the 20 members of the "ensemble."

Combined Numerical Statistical Models: Scripps
Institute of Oceanography Hybrid Coupled Model

This hybrid approach utilizes a detailed ocean model (like the ECMWF, the HOPE model) coupled with a much simpler statistical atmospheric model. The statistical model is used instead of a general circulation model in order to increase the speed (the statistical runs 100 times faster than the general circulation model) and to filter out noise, which it is believed, only degrades the predictive performance of the model.

In the statistical model, historical sea surface temperature patterns are broken into a series of EOF patterns. They decompose the historically observed wind fields in the same way and determine how likely each wind pattern will be associated with each sea surface temperature pattern.

These relationships are applied in the model by applying the wind pattern associated with the model forecast sea surface temperature pattern. The model performs quite well at the 6-month forecast time in the Tropical Pacific. Scripps described the performance at 12 months as "still quite respectable."

Chapter 9

The Super El Niños

THE GREAT EL NIÑO OF 1982–1983

As 1982 began, 5 years had passed since a significant ENSO event, either warm or cold. Mostly neutral conditions had prevailed in the Tropical Pacific. But that was about to change and in a very big way.

ENSO events, both warm and cold, help transfer excess tropical heat poleward (in La Niñas in the western Pacific, in the El Niños further east). This helps compensate for the imbalances of radiation and heat with latitude. If the transport of heat poleward is reduced in any way, the contrasts build and eventually something big happens.

Examples of this are abundant in weather in virtually all seasons and regions. In spring and summer, thunderstorms result from excess heat energy at and near the ground, which makes the atmosphere unstable. This initiates rising air currents and eventually clouds. With continued heating of the ground and air by the sun, at some point, the rising air becomes buoyant enough to rise high enough for showers or thunderstorms to form. In these showers and thunderstorms, the heat is carried upward and rain-cooled air downward to restore equilibrium (stability) to the atmosphere.

Often on a warm spring day in the Plains states, strong sinking air associated with high pressure puts a lid (called a cap) on convection (thunderstorms). The sun warms the ground but the heat energy is trapped in low levels by the sinking air. When an approaching cold front removes the cap, explosive thunderstorms and possibly severe weather occurs.

Similarly, in the Tropical Atlantic, strong sinking associated with the subtropical high-pressure acts to suppress storm formation in the spring and early summer. This acts like a pressure cooker allowing the heat to build in the ocean and lower atmosphere. When the high shifts north later in the summer, tropical waves easily trigger strong thunderstorms in the very warm, moist, and unstable air. These thunderstorms can organize and develop into tropical storms and hurricanes.

In much the same way, along the East Coast of the United States in winter, cold air moving offshore over warm water makes the lower atmosphere very unstable and low clouds quickly form. However, high pressure here also "caps" the clouds. If a low pressure then approaches and the cap is lifted, strong thunderstorms can develop offshore. The sudden introduction of heat into the low pressure can help explosively intensify the storm into the famous "nor'easters," which can bury the major cities of the east under heavy snows.

In the Tropical Pacific, if the heat is allowed to build up in the ocean in the absence of a strong ENSO event, eventually a very strong event may be triggered. This may have been a factor in the Great El Niño of 1982–1983 (Table 9.1). This event began quietly in May 1982, when the easterly winds that normally blow across the equatorial Pacific from off the South American coast to Indonesia, began to weaken. The winds soon shifted to westerly west of the dateline. The ocean reacted quickly to the changes in the winds. Normally, sea levels in the western Pacific are higher than in the eastern areas because of the action of the steady easterly winds. As the easterly winds diminished, the water began to slosh eastward. Sea levels quickly rose several inches in the mid-Pacific by June 1982. By October, sea level rises of up to foot had spread all the way to the coast of South America.

In the eastern Pacific, the ocean temperatures increased due to the combination of weakened upwelling and the arrival of warm water from the west. Off the coast of South America, the water warmed from the low 70s (°F) well into the 80s (°F).

Precipitation, which is normally concentrated in the western equatorial Pacific, shifted eastward with the warm water. Monsoon rains fell over the central Pacific instead of the west. Typhoons and hurricanes feeding off unusually warm water were steered off their normal tracks to islands like Tahiti and Hawaii, normally unaffected by these storms.

Table 9.1
The Effects of the Great El Niño of 1982–1983

Region	Weather	Deaths	Losses
Australia	Drought and fires	71	$2.5 billion
Indonesia	Drought	340	$500 million
Philippines	Drought		$450 million
Southern India, Sri Lanka	Drought		$150 million
Southern China	Flooding	600	$600 million
Hawaii	Hurricane	1	$230 million
Tahiti, French Polynesia	Hurricane, tropical storms	1	$50 million
Southern Brazil, Northern Argentina, East Paraguay	Flooding	170	$3 billion
Bolivia	Flooding	50	$300 million
Southern Peru, Western Bolivia	Drought		$240 million
Ecuador, Northern Peru	Flooding	600	$650 million
Mexico, Central America	Drought		$600 million
Cuba	Flooding	15	$170 million
Southern Africa	Drought		$1 billion
United States	Storm damage	160	$2.2 billion
Northern Africa, Iberian Peninsula	Drought		$200 million
Western Europe	Flooding	25	$200 million
Middle East	Cold, snow	65	$50 million
TOTALS		2098	$14 billion

Source: J. M. Wallace and S. Vogel (1994), *El Niño and Climate Production, Reports to the Nation on Our Changing Planet.* Courtesy of National Oceanic & Atmospheric Administration (NOAA).

In Peru and Ecuador, heavy rains occurred in normally arid regions. More than 100 inches of rain fell in places that average less than 6 inches of rain in a normal year. Meanwhile, in the western Tropical Pacific in Indonesia and Australia, the story was instead drought. The developing drought fed forest fires. Similar drought conditions developed in the other typical El Niño dry regions like southern Africa, southern India and Sri Lanka, southern Peru and western Bolivia. Drought also did damage in parts of northern Africa and the Iberian Peninsula.

During the winter, strong storms smashed ashore in California and then dumped tons of rainfall across all of the southern United States. These storms did damage also in the Caribbean in places like Cuba. In February, one of the storms brought a paralyzing snowstorm to the big cities of the eastern United States. An estimated 49.5 million people were affected by this one storm in a major way, more than 22 million people received more than 20 inches of snow.

The El Niño faded in the spring and by late summer conditions in the Tropical Pacific (both atmosphere and ocean) had returned to normal.

Summary of the Effects of the Great El Niño of 1982–1983

The true legacy of the Super El Niño of 1982–1983 was the spark it gave to scientific research. The result of that research has been a better understanding of ENSO's true global significance. The observing systems and the numerical and statistical models that have been put into place have enabled us to monitor and forecast ENSO. This has led to the first reliable seasonal forecasts as far as a year in advance.

THE SUPER EL NIÑO OF 1997–1998

The Super El Niño of 1997–1998 came on strong at the end of the winter of 1996–1997. By summer, it showed signs of becoming the strongest ever recorded. Although in the end it ranked by most measures slightly less strong than the 1982–1983 event, it was nonetheless by all accounts a Super El Niño.

By some unofficial estimates, the Super El Niño of 1997–1998 killed 2,100 people and caused $33 billion (U.S.) in property damage worldwide. In Peru, rains in places exceeded that of the 1982–1983 and the results were very similar. Rains, some days falling at a rate of 5 or 6 inches per day, produced massive mudslides and floods. Runoff from flooding poured into the coastal Sechura desert. It transformed the desert wasteland into the second largest lake in Peru (90 miles wide, 20 miles long, and 10 feet deep).

Forest fires burned more than 1 million acres of forest land in

Sumatra, Borneo, and Malaysia. The smoke made breathing difficult (even dangerous) and drivers were forced to use headlights at noon. Temperatures reached 108°F as far north as Mongolia; Madagascar was hit with devastating cyclones; Kenya had more than 40 inches more rain than normal; and floods killed 115 people in central Europe.

In China, the worst flooding in more than 44 years occurred during the summer of 1998 at the end of the El Niño. The central and southern parts of the country were battered by more than 60 days of heavy flooding along the banks of the Yangtze River and its tributaries. In July and August, extensive flooding occurred in the northeast along the Songhuajiang, Nenjiang, and other rivers. According to government estimates, 223 million people—one fifth of China's population—were affected; 3,004 died and 15 million were left homeless. The damage exceeded $20 billion (U.S.).

In Australia, the typical El Niño drought took place. Melbourne, Australia experienced its second driest year in 140 years of record. Careful planning for the event helped reduce the damage from bushfires. In April 1998, heavy rains at the end of the event also helped relieve the drought conditions.

In the United States, winter storms caused mudslides and flash floods from California to Mississippi, storms raked the Gulf Coast and record rains and tornadoes ravaged Florida, normally the sunshine state in winter. Eighteen presidentially declared disasters from the fall of 1997 through April 1998 were attributable in part to El Niño's effects on the atmosphere according to Ants Leetma, Climate Prediction Center.

Los Angeles had 31.01 inches (210 percent of normal) and San Francisco 47.19 inches (230 percent of normal) of rain. Santa Barbara, California set an incredible monthly rainfall total in February 1998 of 21.74 inches, the most for any month back to 1867. Tampa received 15.57 inches of rain in December alone and Orlando had 12.63 inches.

During the late evening of February 22 and the early morning of February 23, a large outbreak of tornadoes occurred across Florida with 42 fatalities. Eight hundred homes were destroyed, another 700 were left uninhabitable, and another 3,500 damaged. Some 135,000 utility customers lost power. The damage from the outbreak exceeded $60 million bringing the total for the winter to $500 million. Of Florida's 67 counties, 54 were declared federal disaster areas during the 1997–1998 El Niño winter.

The southern storm track brought unusual snows to some places before the El Niño warmth kicked in. Twenty-one inches of snow fell near Roswell, New Mexico in December, the largest 1-month total since 1893. Snow also fell in December in Guadalajara, Mexico for the first time since 1881.

Drought grew throughout the equatorial climate zones of Central and South America. Fires in Guatemala, Honduras, and Nicaragua consumed 2,150 square miles of rain forest. In Mexico, where the drought was the worst in more than 70 years, 1,500 square miles burned. In Brazil, an incredible 20,000 square miles (an area as large as New York State) was destroyed by flames.

Chapter 10

How Governments Use ENSO-Based Seasonal Forecasts

Because of its proximity to the warm and cold water plumes of El Niño and La Niña, and because of its economy's dependence on fishing and agriculture, the small country of Peru was the first government to carefully monitor the phenomena and use forecasts in public policy and planning. In northern Peru, El Niños are generally wetter than normal. During strong El Niños like those of 1982–1983 and 1997–1998, heavy rains and flooding can occur. During La Niñas, drought can be a problem.

After the record El Niño of 1982–1983, the Peruvian government decided to develop a program to predict future ENSO happenings. At the time, they were concerned that a strong La Niña might follow the next season and bring drought and crop failures.

The first forecast was made in November 1983 for the early 1984 rainy season. At the time, climate conditions were near normal and forecast to remain so. These conditions are generally favorable for agriculture. These forecasts were distributed to several organizations and to the minister of agriculture, who incorporated it into their plans for the growing season. The first forecasts were correct and the harvest was abundant. Since 1983, these forecasts have been prepared each November based on winds and ocean temperatures across the Pacific, and in recent years on numerical and statistical models.

Farmers in northern Peru grow crops such as cotton and rice. Rice thrives on wet conditions during most of the growing season, while the deep-rooted cotton crops tolerate drought better. Farmers usually grow more rice and less cotton in El Niño years and more cotton and less rice in La Niña years.

Other countries followed suit, including Australia, Brazil, Ethiopia, and India. In Australia, the Bureau of Meteorology uses historical relationships between the values and trends of the SOI and precipitation to predict the probabilities of rainfall exceeding median rainfall for five regions in Australia. The forecasts are issued monthly and are used by government and agriculture.

In Brazil, El Niños bring a reduction in rainfall and increased probability of drought. In 1987, no action was taken with a forecast of drought. When a 30 percent shortfall of rainfall occurred, grain production was less than 20 percent of normal. In 1992, when El Niño was predicted, drought-resistant crops were planted. Precipitation was again about 30 percent below normal during the growing season, but grain production was far greater, nearly 80 percent of normal. This is an example of how governments and industry can take advantage of the ENSO forecast.

African nations find extremes of drought and flood associated with both El Niño and La Niña. Flooding in the east African nations of Tanzania, Kenya, Somalia, and Ethiopia during the El Niño of 1997–1998 contributed to a large increase in animal and human diseases. The heavy losses of livestock and trade embargoes on diseased livestock, in turn, affected the welfare and incomes of a large number of people in the region. The Food Health Organization (FHO) worked with other agencies and the veterinary services of the nations involved to help quickly eradicate the diseases there after the El Niño waned.

The Food and Agriculture Organization (FAO) is an example of international agencies that maintain close collaboration with government, donors, and humanitarian agencies involved in alleviating the adverse effects of these ENSO-related climate extremes. They promote a variety of measures such as the following:

1. Support for irrigation projects in South Africa, Central America, and the Caribbean.

2. Development of drought- and storm-resistant cropping, farming, and fishing patterns for South Asia, the Sahel in Africa, eastern and southern Africa, and the Caribbean.

3. Support for a disaster preparedness strategy for the member countries of the Intergovernmental Authority on Development in eastern Africa.

4. Provision of information and technical advice on forestry policy and planning, sound forest management, logging policies, and fire control.

5. Provision of information and technical advice on fishery disruptions causes by changes in ocean temperatures associated with ENSO.

6. Support for flood prevention programs.

Tropical countries have the most to benefit from good ENSO forecasts because they experience a disproportionate share of the impacts from ENSO events and because the climate models used to predict ENSO events are most accurate in this area.

ENSO forecasts help countries anticipate and mitigate droughts and floods, and are very useful in agricultural planning. Countries such as Brazil, Australia, India, Peru, and various African nations that are in latitudes with strong El Niño connections to weather patterns use ENSO predictions to help agricultural producers select crops most likely to be successful in the coming growing season. In countries or regions with a Famine Early Warning System (FEWS) in place, ENSO forecasts can play a key role in mitigating the impacts of flood or drought that can lead to famine. Famine, like drought, is a slow-onset disaster, therefore advanced warning may enable countries to greatly reduce, if not eliminate, its worst impacts.

An example that illustrates how governments have successfully used seasonal forecasts based on ENSO occurred in northeastern Brazil in the early 1990s. The region had suffered through an extended period of drought. The Brazilian city of Fortaleza (a city of several million people) would entirely deplete its water resources if the drought continued for another warm season. The Brazilian government consulted with climate forecasters at Scripps Institute of Oceanography. They predicted another drought year. The Brazilian government then invested in building a canal to Fortaleza. As predicted, the rains failed but the canal was finished just in time. It saved millions of dollars in damage and protected public health.

In the United States, state governments and agencies in areas most threatened by ENSO instituted their first mitigation efforts during the 1997–1998 El Niño. For example, California took seriously the warnings that storms and flooding would be similar to the Great El Niño of 1982–1983, which produced about $2 billion in damages. The state of California thus spent $7.5 million to aid in preparedness and alert the public. Several communities spent their funds on local projects designed specifically to minimize damage and loss of life. Homeowners also spent an additional $125 million on home repair and roofing in advance of the winter storms (Labor Department, 1998).

Then, during the winter of 1997–1998, California was indeed assaulted by powerful Pacific storms that delivered copious rainfall, and caused a major flooding, landslides, and losses to agriculture that totaled $1.1 billion (Changnon, 1999). Although it is not possible to estimate precisely how much was actually saved by the mitigation efforts, it was generally agreed that the efforts by the state and local agencies saved many lives and limited property loss since the total losses were much less than in 1982–1983.

CALIFORNIA RESPONSES TO 1997–1998 EL NIÑO

The 1997–1998 California statistics compiled by FEMA illustrate the actions taken and their results.

City of Oakland

Following the El Niño Community Preparedness Summit in Santa Monica in October 1997, the city of Oakland, a Project Impact community, began organizing its approach to defend itself against predicted El Niño-generated rains. The city's Department of Public Works headed the El Niño response. Project Impact is a FEMA initiative that helps build disaster-resistant communities. The city committed $3 million to preventive measures, including clearing hillside spillways of brush and debris to ensure natural flow into storm water channels. When not cleared, debris and brush divert water away from proper channels, causing erosion and flooding. Lake Merritt is a tidal lake and in 1997 the combination of high tides and heavy rains caused the lake to flood nearby businesses. Sandbags were used to keep the lake from flooding stores. The city distributed sandbags to residents through fire stations (a total of 110,000, about 20 per household). The city delivered sandbags directly to seniors and individuals with disabilities. The city also instituted an adopt-a-drain program in which individuals voluntarily kept drains clear of debris during heavy rains, thereby facilitating better drainage and minimizing erosion-causing runoff. The city issued rain slickers, hats, and rakes to all volunteers. During the El Niño rains, the city identified critical areas subject to flooding and directed the fire department to patrol those areas and report potential problems to a centralized emergency response center. Public Works, Parks & Recreation, and Building Services worked together to quickly remove fallen trees to keep hillside runoff flowing efficiently. The city also instituted a 24-hour telephone line giving residents around-the-clock ability to report problems. Deputy Chief Donald Parker of the Oakland Fire Department said, "Damage would have been a lot worse if we hadn't prepared."

Los Angeles County

Los Angeles County began planning for El Niño soon after the 1997 winter floods. A special task force included the following county departments: Emergency Services, Public Works, Fire, Sheriff, Health Services, Internal Services, Public Social Services, and Coroner. Also included on the task force were representatives from Civil Defense and the American Red Cross. The task force developed a plan based on FEMA's definition of a 100-year flood causing the Los Angeles River and Rio Hondo Flood Channel to overflow.

Some of the actions taken by the county included the following:

- Distributing sand and sandbags to high-impact areas.
- Establishing an El Niño Web site.
- Cleaning drains and flood channels, and bolstering levees by mid-August.
- Providing special training for damage-control teams.

In addition to mitigation efforts, Los Angeles County also honed its emergency responses, including the review of evacuation needs, location of possible shelters, and ongoing emergency response training in all county departments.

Colusa County

Several homes and commercial buildings in the city of Colusa flooded during the 1997 floods. The main source of flooding came from water that drained across agricultural fields, overflowed a drainage ditch, and continued into town. During heavy rains in early February, the city learned of a levee that had broken several miles from town. Surveys taken by the community before the rainy season identified the route drainage water would flow. The county petitioned the California Department of Water Resources for sufficient funding to erect a 3-foot high berm on the west side of the city, less than 1 mile in length. Within hours, the berm was constructed. The berm did its job by containing the water in the field just yards from 22 threatened structures.

City of Manteca

The Wetherbee Lake area of Manteca flooded in 1997 when levees broke during heavy runoff from warm rains and snowmelt. The U.S. Army Corps of Engineers completed an 11-month levee repair project just as the El Niño storms arrived. The levee held and residents were spared from a second flooding. One resident who was flooded in 1997 rebuilt his home, elevating it 12 feet, 4 feet above the base-flood elevation.

Monterey County Farmers

Farmers and landowners along the Salinas River in Monterey County banded together to reduce flooding that caused $240 million in damages in 1995. They formed a coalition, spent $2 million to clean out vegetation, sandbars and other flow impediments along 40 miles of the river, thereby increasing the water flow capacity of the river by 33 percent. The result was that the Salinas River did not flood during the El Niño '98 storms.

Guerneville

Many people living in areas close to rivers prone to flooding elevated their homes. In Guerneville on the Russian River, a community effort raised funds for materials needed to elevate 14 homes above base-flood level. Volunteers did the heavy work of elevating the homes and the FEMA Hazard Mitigation Grant Program provided funds to raise another 200 homes.

Seal Beach

In Seal Beach, south of Long Beach, city and Orange County crews, in anticipation of El Niño-driven pounding surf and high tides, built a 10-foot high berm several hundred yards long on the beach to protect scores of beach-front homes.

Orange County

Other mitigation measures taken by Orange County included cleaning and monitoring of storm drains, stockpiling sand and sandbags, and constructing sand berms to protect beachfront property.

St. Helena

St. Helena has a history of repetitive flooding. In anticipation of El Niño, the city cleared debris from the Napa River and established a close monitoring of the tributary. This was a concerted effort of business and residents working together. The results were successful during the 1997–1998 El Niño storm season.

San Leandro

When El Niño rains saturated the ground in the San Francisco Bay Area, an entire hill in San Leandro began to slide, threatening several homes. The city financed the moving of three homes that were in immediate danger, and continued to monitor the hill that threatened four more homes. The decision to move the three homes was based on the fact that it was less costly to move a home than to rebuild.

The Entire State of California

People statewide responded to El Niño warnings. The National Flood Insurance Program reported a surge in Californians purchasing flood insurance following the El Niño Community Preparedness Summit. The number of policies went from a pre-summit total of 264,914 to 333,753 by the end of November 1997. The number climbed to 365,000 by the end of December 1997.

THE INTERNATIONAL RESEARCH INSTITUTE (IRI)

Under a cooperative agreement with the U.S. National Oceanic and Atmospheric Administration's Office of Global Programs, Lamont-Doherty Earth Observatory at Columbia University and Scripps Institute of Oceanography were asked to establish an International Research Institute (IRI). IRI's mission was to provide experimental forecast guidance on a seasonal basis to communities around the world that are significantly affected by ENSO. The IRI provides assessments of the expected conditions and the possible regional consequences to critical economic sectors such as agriculture, water resources, fishing, emergency preparedness, and public health and safety.

It is an end-to-end program that goes well beyond just seasonal or interannual climate prediction. The IRI transforms emerging seasonal climate forecasts into useful information to support operational decision making.

The IRI also provides a valuable bridge between the scientific community engaged in the exploration of climate prediction and the principle beneficiaries of these forecasts and insights. The services are most needed by nations in the tropics where most of the world's people reside. Fortunately, that is where correlations of climate with ENSO are best and climate forecasts most accurate.

The IRI has implemented a program consisting of modeling research, state-of-the-art forecasting, applications research, climate monitoring, information dissemination, and education and training. (For additional information contact the Office of Director, International Research Institute, Lamont-Doherty Earth Observatory of Columbia University, 204 Oceanography Building, Route 9W, Palisades, NY 10964, Phone: 914 365–8368, http://inpred.idgo.columbia. edu/.)

NOAA ESTIMATES OF ENSO EFFECTS ON ECONOMIC ACTIVITIES

NOAA has estimated that by incorporating ENSO forecasts into planting decisions, farmers in the United States could increase agricultural output and produce benefits of $320 million per year. The forecasts can help them make decisions about which crops and hybrid varieties to plant. For example, if drier than normal conditions are expected during a growing season, they can plant more drought-resistant varieties, thus improving crop yields.

For fisheries, ENSO affects both water temperature and streamflow in spawning areas, which in turn can have an effect on fish populations and reproductive behavior. For example, in the one small northwestern U.S. coho salmon fishery, use of ENSO forecasts have been estimated to have produced benefits of nearly $1 million per year or about 10 percent of the landed value produced by the fishery.

For energy-distribution companies, a forecast of a colder than normal winter (as might be expected in the northwest or north central states in a La Niña), might lead to an increase stockpiles of heating oil and natural gas, thus avoiding the costly shortfalls. Likewise, a forecast of a warm winter as in a strong El Niño, might lead to stockpiling less and drawing down inventories faster.

Further upstream in the supply chain, at the refinery level, forecasts of warm or cold weather due to ENSO events may cause producers to change the output mix of gasoline and heating oil. In other words, increase the production of heating oil in the anticipation of a cold winter, or produce less heating oil and more gasoline in a warm one.

Homeowners and businesses in areas where El Niño storms often hit hard (like California and Florida) may contract with roofing contractors to reduce the risk or extent of damage to the inside of structures due to leaky roofs in winter storms. More generally, in the construction industry, accurate long-term forecasts based on ENSO may help managers better plan and schedule projects that are weather dependent.

For water supply managers, ENSO forecasts can help with decisions about water storage in reservoirs or man-made lakes behind hydroelectric plants. For example, they might carry less water in storage in advance of an expected wet season, and tend to use more water early in the season. Conversely, in a season that is forecast to be dry, they would tend to store more and use less early in the season in anticipation of the drier conditions.

Table 10.1 summarizes the effects of ENSO on climate-sensitive industries. The industries listed (agriculture, energy distribution, water-supply management, recreation and construction account for nearly 15 percent of the gross domestic product).

Economic Losses and Benefits from El Niño

Clearly then, ENSO events are capable of producing losses that can total in the billions of dollars. El Niños are feared in places like Indonesia, Australia, India, Brazil, Mexico, and parts of Africa where devastating droughts are possible. In the United States, states from California to Florida are vulnerable to damage from a barrage of strong storms.

In the United States, the El Niño of 1997–1998 played a role in 18 presidentially declared disasters with damage exceeding $4 billion. On the other hand, the patterns of weather associated with strong El Niño events also produce many positive effects. For example, the milder temperatures of strong El Niño winters in the interior northern United States reduces heating costs for both homes and industry and the operating costs for transportation both by air and on land. Less snowfall in the north lowers the costs of snow removal for government and industry, and enables the construction industry to work

Table 10.1
Effects of ENSO on Climate-Sensitive Industries

Economic Activity	Economic Scale of Activity	Effect of Long-Term Weather Fluctuations	How Forecasts Can Be Used
Crop agriculture	$109 billion (1996 U.S. cash receipts)	Temperature and rainfall affect crop yields	Farmers can select crop varieties appropriate to expected temperature and rainfall conditions; distributors can reduce commodity storage if uncertainty about future yields is reduced
Fisheries	$3.5 billion (U.S. landings)	Water temperature and streamflow affect fish populations and reproductive behavior	Fishery managers can adjust harvesting to ensure adequate spawning
Oil and gas distribution	$76 billion (1992 natural gas production and distribution) $7 billion (residential and commercial heating gas and fuel oil, average)	Temperature affects demand for heating fuels	Energy suppliers can adjust fuel stores and better time drawdown of stored fuel
Water supply management	Unknown	Precipitation affects the amount of water entering reservoirs and the demand for irrigation	Water supply managers can improve reservoir management by anticipating future inflows

112

Storm damage mitigation and repair	$16.7 billion (1992 value of roofing/siding construction work)	Storms (wind and precipitation) cause damage to buildings and other infrastructure	Homeowners can take measures to minimize storm damage (preemptive repairs); municipalities can prepare for possible floods (clearing storm drains, drainage channels, etc.)
Recreation	$100 billion (1992 hotel and recreational amusement centers)	Temperature and snowfall affect winter sports; rainfall affects other outdoor recreation	Vacationers can improve their vacation experience by better planning their travel and sports activities
Construction	$528 billion (1992 construction industries)	Temperature and precipitation affect whether construction can proceed	Construction managers can better schedule projects

Source: From NOAA report, *Improving El Niño Forecasting: The Potential Economic Benefits.*

Table 10.2
Economic Gains and Losses during the Great El Niño of 1997–1998

Economic Impact	Losses	Benefits
Property losses	$2,800,000,000	
Federal government relief	$ 400,000,000	
State assistance	$ 125,000,000	
Agricultural effects	$ 675,000,000	
Reduced sales of snow removal equipment	$ 70,000,000	
Tourism, recreation	$ 190,000,000	
Savings heating costs		$6,700,000,000
Increased sales—homes, goods		$5,600,000,000
Reduction in costs of removal of ice and snow from streets and highways		$ 375,000,000
Reduction in losses due to the absence of snowmelt floods and hurricanes		$6,900,000,000
Income from increased construction and related employment		$ 475,000,000
Reduced operating costs to airlines and trucking		$ 167,500,000
TOTAL	$4,350,000,000	$19,750,000,000

Source: S. A. Changnon (1999), Impacts of the 1997–98 El Niño-generated weather in the United States, *Bulletin of the American Meteorological Society* 80, no. 9, pp. 1819–1827.

more during the winter months. Shoppers are able to get to and from stores more easily and more often and retail sales benefit. El Niño also typically results in less flooding during the spring and fewer hurricanes in the summer.

For example, Changnon (1999) estimated the economic gains and losses during the great El Niño of 1997–1998. He showed that the economic benefits may be much greater than the losses (See Table 10.2).

Ironically, in weaker El Niños, the picture may be very different. Weak El Niño winters produce an increased number of storms but not as much added warmth. The result is that there can be heavy snowstorms instead of flooding rains in the big cities of the east, with major (billion dollar) impact on the economy. The cold and snow further reduces or even eliminates many of the benefits that occur in strong El Niños.

Economic Losses and Benefits from La Niña

In La Niña, the picture is very different from that of El Niño. When periodic outbreaks of extreme cold weather and snow occur across the northern states, the costs of heating, snow removal, fuel for airline and trucking industries can become significant. If ice storms occur across the south or east, business may be shut down for days causing major effects on commerce. Retail sales may be down due to travel difficulties, and construction work may be hampered resulting in delays and loss of employment.

Also in La Niñas, losses from springtime flooding and from summer droughts and hurricanes typically are typically much greater than normal. Flooding in La Niña years averages nearly $4.5 billion, compared to an average of $2.4 billion. Hurricane-related losses in La Niña years average $5.9 billion, compared to an average of $3 billion.

Take for example the 1998–1999 La Niña. Hurricanes Bonnie, Georges, Dennis, Floyd, Irene, and Harvey made landfall in the two summers with well over $13 billion in damages. Major tornado outbreaks occurred in January 1999 in Arkansas and Tennessee and in May in Oklahoma and Kansas with $2.3 billion in total damages. The summer of 1999's heat wave and drought in the east central states added more than $1 billion in losses. Losses from La Niña-related storms and lack of storms during 1998 and 1999 exceeded $16 billion.

A heat wave and drought in 1996 was responsible for $5 billion in losses in the south central states. A major heat wave and drought in 1988 caused an estimated $40 billion in damage or losses (mostly agricultural) in the central and eastern United States.

On the other hand, the winter sports industry benefits in the west and north from increased snowfall. Tourism in "escape" destinations like Florida and California usually increases. Sales of snow removal equipment and winter clothing are also higher. But the benefits may be dwarfed by the losses.

Ironically, at least in the United States, La Niñas can be much more costly than El Niños, despite the fact that most of the media attention is on El Niño.

Chapter 11

ENSO, Tip of the Iceberg in Climate Prediction

ENSO research, sparked by the great El Niño of 1982–1983, has uncovered strong global correlations between ENSO and climate. This has enabled skillful, multiseason outlooks. The success of the ENSO correlations sparked research into possible climate relationships with other oscillations and factors. Some of these other relationships have shown great promise. These outlooks may help researchers refine ENSO forecasts during events and forecast the climate even during neutral ENSO years.

NEUTRAL YEARS AFTER ENSO

The oscillation that is ENSO results in a flip-flop between cold (La Niña) and warm (El Niño) conditions in the eastern Tropical Pacific. Sometimes the transition is a rapid one lasting just weeks. This was the case in 1998 when strong El Niño conditions turned into strong La Niña conditions in less than 2 months. Sometimes the transition is a very gradual one, resulting in a year or more in a neutral state. In the last half-century, we have been in a neutral state in 17 (34 percent) of the winters.

The neutral state really doesn't have a name, so we are giving it one—La Nada (literally "the nothing"). There are some reliable features of both El Niño and La Niña but are there any common patterns in some or all La Nada years?

The Atmosphere and Ocean Apparently Have a Memory

As we can see in plates 5, 6 and 7, and there is a tendency for La Niña-like patterns to occur in neutral ENSO (La Nada) years following La Niña and El Niño-like patterns in years following El Niño. Presumably only during neutral years following neutral years are the patterns truly random or controlled entirely by other factors.

La Nada conditions in the years after El Niño or La Niña show some continuity of the pattern. In other words, a hint of the El Niño after El Niño and a hint of La Niña after La Niña.

INTERACTION OF ENSO WITH PRESSURE PATTERNS IN THE ATLANTIC

Weather patterns in the North Atlantic are subject to an oscillation not unlike ENSO. This North Atlantic version, not surprisingly, is called the North Atlantic Oscillation (NAO).

The Southern Oscillation involves a flip-flop in relative strength of pressure systems from the western to eastern Tropical Pacific. The NAO involves a flip-flop in relative strength of pressure systems north to south in the North Atlantic. Normally, on average in the North Atlantic in winter, low pressure is found in the north near Iceland (called the Icelandic low) with high pressure to the south off of Portugal or the Azores (called the Azores High).

At times, the Icelandic low and/or the Azores high become especially strong resulting in a very fast jet stream flow across the Atlantic and into western Europe. This drains cold air off of North America and often results in above normal temperatures especially in the southeastern states. The air is warmed and moistened over the relatively warm Atlantic waters then the westerly flow off the North Atlantic into Europe carries this mild, moist, maritime air inland resulting in cloudy, wet, and mild conditions (Figure 11.1).

Occasionally, this pattern flip-flops with high pressure developing or moving into the far northern Atlantic (often settling temporarily into a position near Greenland, where it is called the Greenland block). Meanwhile, relative low pressure develops to the south displacing the Azores high. This negative phase of the NAO often results in harsh winter weather over eastern North America and in Europe. This is because it retards the passage of cold air off of North America

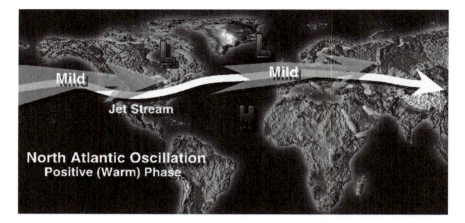

Figure 11.1.   The positive mode of the NAO. This pattern produces the mildest weather in both the United States and Europe.
Source: Intellicast Dr. Dewpoint—http://www.intellicast.com/DrDewpoint/ Library/1137/2/
Courtesy of Intellicast.com

(thus the term blocking). The cold air pools and expands well to the south over the eastern United States. The storm track too is pushed south of normal often resulting in snowstorms for the eastern metropolitan areas (Figure 11.2).

In Europe, the easterly steering flow beneath the North Atlantic blocking highs often draws cold air from Siberia west to replace the normal mild maritime air over western Europe. Snow can accompany the unusual frigid winter temperatures. Cold and snow in Europe usually signify that the NAO has turned negative. These blocks are usually transitory and last on average a week or two. If conditions are right, the blocks redevelop again after a period of time. In some years, (such as 1995–1996), the negative phase of the NAO can repeat itself frequently. In Figures 11.3 and 11.4, we compare snowfall in 1995–1996 to snowfall in the winter of 1988–1989 when the NAO was strongly positive.

It appears the NAO operates independently of ENSO, as both modes of the NAO are present in El Niño and La Niña. However, the NAO can significantly influence the weather in both El Niño and La Niña years. A negative NAO combined with either El Niño or La Niña can greatly magnify the cooling effects in both Europe and the eastern United States. In all but the strongest El Niños, a negative

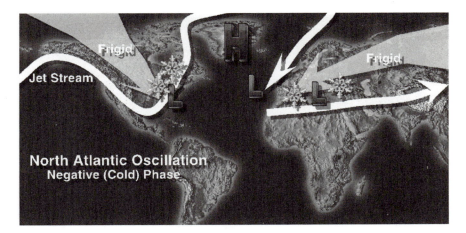

Figure 11.2.   The negative mode of the NAO. This pattern produces the harshest winter weather in both the United States and Europe.
Source: Intellicast Dr. Dewpoint—http://www.intellicast.com/DrDewpoint/Library/1137/2/
Courtesy of Intellicast.com

NAO will cause the trough in the eastern United States favored during El Niños, to be particularly strong. This helps draw down more cold air and produce more eastern and southern snowstorms. Examples of this are the very snowy years of 1957–1958, 1968–1969, 1977–1978, 1987–1988. In La Niña years, a negative NAO bottles up cold air over North America and depresses the storm track normally across the far northern United States, south. These storms then can bring frequent snows to the major cities of the northwest, midwest, and east.

In Europe, a negative NAO can help draw cold air from the arctic or from eastern Europe westward to normally mild areas of western Europe in both El Niño and La Niña years.

In the 1999–2000 winter outlook, the Climate Prediction Center looked at the temperature, rainfall, and snowfall patterns in La Niña winters with the two modes of the NAO. The analysis shows this same very interesting result. Cold and snow is far more abundant and further south in winters with a net negative NAO than in winters with a net positive NAO.

Longer term, the NAO seems to go through a decadal scale cycle. It was predominantly in a negative mode from the late 1930s through

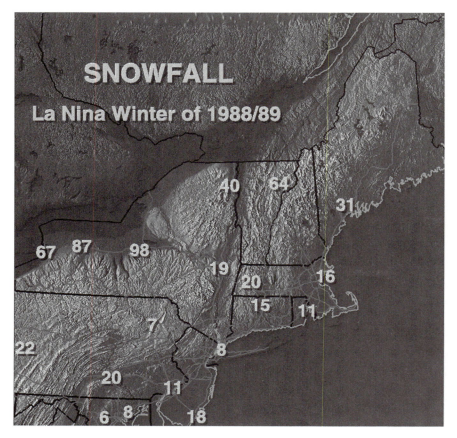

Figure 11.3.   Seasonal snowfall in the La Niña winter of 1988–1989, when the NAO was a high +5. Note the lack of snow even across the northern areas of New York and New England. Only the lake snow belt had appreciable (although below normal) snows.
Source: Intellicast Dr. Dewpoint—http://www.intellicast.com/DrDewpoint/ Library/1137/2/
Courtesy of Intellicast.com

the early 1970s, then predominantly positive from the 1970s to the mid-1990s (see Figure 11.5).

As is the case for the phases of ENSO, these longer term tendencies may be related to the development of warm and cold pools of ocean water. In the part of the cycle where the negative phase dominates, waters tend to be warmer than normal in the tropical Atlantic and in the far north Atlantic and cooler than normal in between.

This is believed to be related to changes in the thermohaline cir-

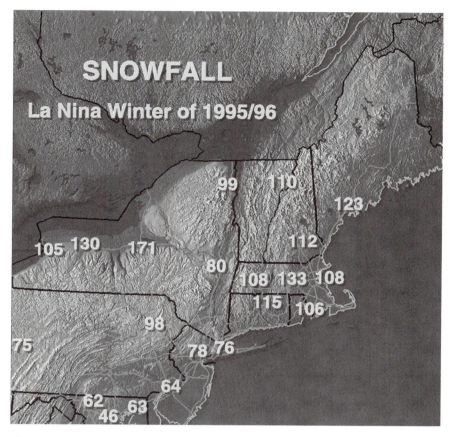

Figure 11.4.   In sharp contrast note the increadible snow totals in the all-time record winter of 1995–1996 in the northeast. In this winter, the NAO was strongly negative (-4, only 1968/69 was more negative). Above normal snows accompanied blocking periods in every month from November to April.
Source: Intellicast Dr. Dewpoint—http://www.intellicast.com/DrDewpoint/ Library/1137/2/
Courtesy of Intellicast.com

culation in the Atlantic. This circulation starts in the polar region where cold, relatively fresh water sinks and is replaced by warmer surface water from the south. In the negative phase, this circulation increases, allowing more warm water to reach the far north. In the positive phase, the circulation slows and cold water accumulates in the far north (see Plate 8a).

Dr. William Gray (1984) in his April update to the Hurricane Sea-

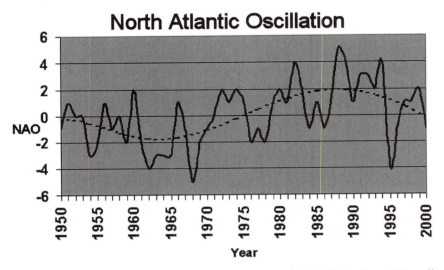

Figure 11.5.   Winter (December through March) NAO Indices (Hurrell, 1995). The winter NAO was predominantly positive in the early 1900s, predominantly negative from the late 1930s to the early 1940s then positive from the 1970s to the mid-1990s. The NAO turned strongly negative again in 1995/96, perhaps signaling another long-term shift in NAO tendency.

Source: University of Washington—http://tao.atmos.washington.edu/data_sets/ nao/ JISCO, Gudmonson, Candace; gcg@atmos.washington.edu; editor; office: 310B ATG; 206–685–9529

Courtesy of University of Washington (Hurrell, 1995)

son 2001 (Web site forecast and press release from Colorado State University). Outlook discusses the evidence for this thermohaline circulation change:

> Three decades have passed since these SST anomaly patterns have been this warm. During June through September 1999 SSTA values in the North Atlantic (50–60°N, 10–50°W) were nearly 1°C warmer than in the earlier five-year (1990–1994) period. These warmer SSTAs are presumably a result of a stronger Atlantic Ocean thermohaline circulation which has also led to a 0.5°C warming of the tropical Atlantic (6–22°N, 18–50°W).
>
> Recent observations indicate increased salinity in upper layers of the North Atlantic. Greater salinity increases water density of these surface layers which are then more able to sink to great depth, thereby increasing compensating northward flow of warm (and salty) replacement water at upper ocean levels. The result-

ing net northward transport of upper-layer warm water into the high North Atlantic (and compensating equatorward transport of deep cold water) is the principal manifestation of the Atlantic Ocean thermohaline ("Conveyor") circulation. A strong conveyor circulation transports greater quantities of heat to high latitudes. Hence, slowly rising salinity values in the far North Atlantic during recent years indicates the development of a stronger thermohaline circulation and a warmer North Atlantic. The effects of a stronger thermohaline circulation have been evident in the region since the spring of 1995 where, as noted before, the best proxy for this increased circulation has been warm North Atlantic SST anomalies.

ENSO AND THE PACIFIC NORTH AMERICAN INDEX

The PNA Index is a measure of one of the primary modes of global climate variability. It measures pressures relative to normal in the upper atmosphere at jet stream level across a large area from the eastern Pacific to eastern North America. This steering flow pattern and associated surface weather features vary day to day. However, because of the different thermal characteristics of the land and ocean, the large north to south mountain barrier(s) that interfere with the flow pattern and the earth's rotation on its axis, the upper flow tends to conform to one of two general phases (the two modes of the PNA).

In the positive phase of the PNA, (Figure 11.6) pressures are below normal and the jet stream dips to the south in the eastern Pacific, then turns north as it passes over the western mountains and then turns south again east of the Rocky Mountains. When it is amplified in winter, this mode can deliver very cold air to the eastern two thirds of the nation. This mode is favored during El Niño.

In the negative phase of the PNA, (see Figure 11.7) pressures are above normal off the west coast pushing the jet stream north of its normal position. Pressures then tend to be below normal in the western United States with the polar jet stream depressed south of normal. The jet stream then turns north, east of the Rocky Mountains. This pattern brings the coldest weather to the western mountains and mild conditions to the southeastern states. This mode is favored during La Niña winters.

The PNA changes phase periodically during a typical season, even when one phase dominates. In the mean however, as the following

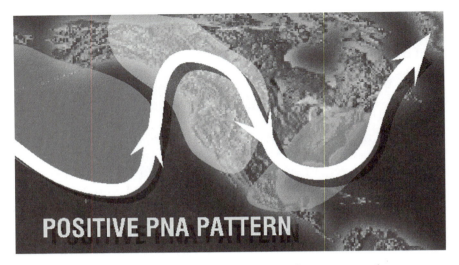

Figure 11.6.   Positive PNA Phase: above normal pressures and temperatures in the western part of North America, with below normal pressures
and temperatures to the west in the eastern Pacific and to the east in the
southeastern United States.
Source: Climate Diagnostics Center Map Room Weather Products—*http://www.*
cdc.noaa.gov/~gtb/tele/pna.cmp.gif
Courtesy of Climate Prediction Center

data show, the positive phase is typical during El Niño winters and
the negative during La Niña winters.

El Niño Years and the PNA

Year	Mean Winter PNA
1982–1983	+1.0
1997–1998	+1.0
1972–1973	−0.3
1991–1992	+0.7
1965–1966	−0.6
1957–1958	+0.3
1968–1969	−0.2
1986–1987	+0.7
1976–1977	+1.0
1977–1978	+0.4
1994–1995	+0.0
1953–1954	+0.6
1969–1970	+0.3

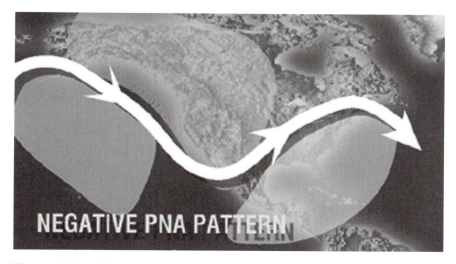

Figure 11.7. Negative PNA Phase: below normal pressures and temperatures in the western part of North America and above normal pressures and temperatures in the southeastern United States.
Source: Climate Diagnostics Center Map Room Weather Products—http://www. cdc.noaa.gov/~gtb/tele/pna.cmp.gif
Courtesy of Climate Prediction Center

La Niña Years and the PNA

Year	Mean Winter PNA
1988–1989	−0.6
1973–1974	−0.1
1970–1971	−0.5
1975–1976	−0.1
1949–1950	−0.4
1955–1956	−1.1
1998–1999	−0.1
1954–1955	−0.4
1984–1985	−0.4
1966–1967	−1.0
1995–1996	−0.1
1950–1951	−0.4
1964–1965	−1.1

THE QUASI-BIENNIAL OSCILLATION

At very high levels of the atmosphere (30 to 50mb or 10 to 12 miles), the winds tend to oscillate from westerly to easterly and then back to westerly on a cycle that averages about 29 months. In a series of papers in the late 1980s and early 1990s, CPC and NCAR researchers found some possible correlations of ENSO winter weather patterns with phases of the Quasi-biennial Oscillation (QBO). It is in winter that the polar vortex and jet streams are farthest south and most sensitive to changes in the equatorial region.

In El Niño years, a westerly QBO usually leads to a stronger net positive PNA pattern, which can result in colder weather and increased risk of east coast storms. In the strongest El Niños, the huge area of warm "bath water" in the eastern Tropical Pacific produces enough warming of the middle and high atmosphere to offset the tendency for cold weather in the eastern United States. The storms may then bring mostly rain to the region.

In El Niño years, an easterly QBO usually means a weaker PNA. Strong El Niño winters with an easterly QBO can be especially mild and snowless, unless the NAO is in its negative mode.

In La Niña years, a westerly QBO likewise seems to lead to a stronger negative PNA pattern with colder weather in the west and warmer weather in the southeast than those with an easterly QBO. In the easterly QBO, La Niña winter on the other hand, colder air tends to be a more frequent visitor to the central states.

The QBO may also affect the NAO tendency, especially in La Niña winters. When the QBO is westerly in La Niña winters, there is a tendency for the NAO to be positive. In easterly QBO La Niñas, a negative NAO and more high latitude blocking is apparently favored.

Researchers at NCAR and the CPC (Labitzke & Van Loon, 1989) also found apparent relationships between winter weather patterns and the QBO and the 11-year solar cycle. Around the peak of the solar cycle and during the quiet years, patterns vary depending on the phase of the QBO. In most areas, the patterns in the westerly phase appear to be nearly the opposite of the favored patterns in the easterly phase. And in most areas, patterns near the solar minimum appear to be nearly the opposite of those at the solar maximum.

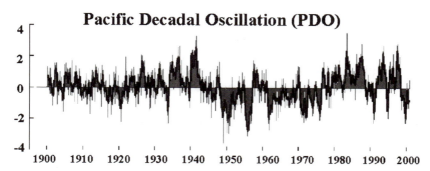

Figure 11.8.   The warm phase predominated during the years from the 1920s to about 1947. The cold phaase followed from 1947 to 1977. The subsequent warm phase then dominated until about 1997. A cold phase existed from 1997 to 2000, perhaps signaling another long-term shift.
Source: http://jisao.washington.edu/pdo/ JISAO; Gudmonson, Candace; gcg@atmos.washington.edu; editor; office: 310B ATG; 206–685–9529
Image courtesy of the University of Washington

THE PACIFIC DECADAL OSCILLATION (PDO)

There appears to be ocean scale oscillations within both the Atlantic and the Pacific that are responsible for major decadal shifts in weather patterns. These may relate to thermohaline circulations that periodically speed up and slow down and control the positions of the large warm and cool water pools. In the Atlantic, this oceanic oscillation controls the favored mode of the NAO. In the Pacific, this oscillation may control the favored mode of ENSO (see Plate 8, bottom).

There is evidence that the PDO has again changed phase in the last few years. Scientists at the Jet Propulsion Laboratory in Pasadena who named this oscillation in the Pacific, the Pacific Decadal Oscillation, say historical records suggest this dramatic shift could last for decades and have a major impact on weather patterns, and both fishing and agriculture.

They claim that satellite imagery shows that the eastern Pacific entered a cool water phase in 1998 that was similar to what was observed several decades ago. Historical data suggests warmer than normal water dominated in the eastern Pacific from 1925 to 1945. A cool phase then developed and lasted from 1947 to 1977, followed by a warm phase from 1977 to 1997 (see Figure 11.8).

Table 11.1
Phases of the PDO, 1947–2000

Phases of the PDO	El Niños	La Niñas
"COLD" phase 1947–1977	1947, 1957, 1963, 1965, 1969, 1970	1949, 1950, 1954, 1955, 1962, 1964, 1967, 1970, 1971, 1973, 1974, 1975
"WARM" phase 1977–1997	1976, 1977, 1982, 1986, 1987, 1991, 1992, 1994, 1997	1984, 1985, 1988, 1995
"COLD" phase 1997-		1998, 1999, 2000

A cool phase favors more La Niñas and an increased risk of summer heat waves and droughts. The cooling of the water, however, may boost fish populations in the Pacific northwest. This is especially true for the salmon populations, which had dropped dramatically after the warming in 1977.

In the warm phase, El Niños are more favored, as was the case over the two decades ending in 1997. Table 11.1 suggests the frequency of La Niñas in the cold phase is about double that of El Niños, whereas in the warm phase, the frequency of the El Niño is double that of the La Niña.

In addition to affecting the frequency of El Niño and La Niña events, the large-scale readjustment to the thermal structure or the Pacific Basin appears to affect how the events evolve. It appears that El Niño events show themselves first in the central Pacific during the warm PDO phases and in the eastern Pacific during the cold phases. Climatologists in the Climate Analysis section of the NCAR have developed the TNI, which considers the relative difference in changes in sea surface temperature from the central to eastern tropical. The TNI changed character after the 1976–1977 winter (with the change in the PDO).

It will be very interesting to see if indeed the next El Niño (due in 2002) shows itself first in the eastern Pacific as the El Niños did in the last cold phase from 1947 to 1977.

Chapter 12

Biographical Sketches

JACOB BJERKNES: November 2, 1897–July 7, 1975

Even if Jacob (known in the United States as Jack) Bjerknes (pronounced B'yerk'nes) had not made the breakthrough connection between the warm Pacific water temperatures associated with El Niño, and the atmospheric pressure shifts associated with the Southern Oscillation, he would have been famous as one of the founders of the modern science of meteorology.

Born in Stockholm, Sweden into a family with a scientific bent, he was the son of Vilhelm Bjernes, also a famous meteorologist, and the grandson of Carl Bjerknes, a Norwegian mathematician who made advances in theories of fluid dynamics. In 1918, when his father Vilhelm began working as a professor at the newly founded Geophysical Institute of the Bergen Museum in Norway, Bjerknes joined him there as a meteorologist, having already served as a research assistant to his father for 3 years at Germany's University of Leipzig.

Together with a handful of other notable meteorologists working there at the time, the father and son team pioneered modern understanding of cyclones, by developing what is known as "frontal theory," with the concept of "warm fronts" and "cold fronts" as boundaries between differing air masses. In 1919, Bjerknes summarized this theory in an eight-page landmark paper entitled, "On the Structure of Moving Cyclones."

In July 1928, Bjerknes married Hedvig Borthen, and the couple

Figure 12.1.   Jacob (Jack) Bjerknes
Courtesy of Eugene Rasmussen, University of Maryland.

had two children. In 1939, while Bjerknes was on an American lec-
ture tour with his family, World War II broke out in Europe, and
because the Nazis occupied Norway, the family was unable to return
to Bergen. Bjerknes remained in the United States, subsequently becoming chairman of the new Department of Meteorology at the University of California at Los Angeles (UCLA), where he worked until
his retirement in 1965.

While at UCLA, Bjerknes made the vital connection between El
Niño and the Southern Oscillation, determining that the Southern
Oscillation was linked to the diminished trade winds that preceded
the onset of El Niño. He was also responsible for coining two terms,
which are in common usage today within the meteorological community. The first is *teleconnections*, which he used to describe the
atmospheric links between the tropics and the higher latitudes. The

second was the *Walker Circulation,* which he named in honor of Sir Gilbert Walker, who years before had first discovered the periodic pressure and temperature fluctuations that occur from one side of the tropical Pacific to the other. Bjerknes' Walker Circulation refers to the three-dimensional circulation in the tropical Pacific associated with El Niño and La Niña. In El Niño the air sinks in the western areas moves eastward and rises in the central (and in stronger El Niños the eastern) Pacific only to return westward at higher altitudes. In La Niña, the pattern reverses as air sinks in the eastern Pacific and moves westward to rise in the western Pacific.

MARK A. CANE: October 20, 1944–

Dr. Mark A. Cane is a professor in the Department of Earth and Climate Sciences at Columbia University's Lamont Doherty Earth Observatory, in Palisades, New York. Together with Lamont colleague and former student Dr. Stephen E. Zebiak, he made a "pivotal breakthrough" (Stevens, 1996) by devising the first numerical model simulating El Niño and the Southern Oscillation. In 1985, this model was used to make the first physically based forecasts of El Niño, and it has been used extensively by many investigators since that time.

Dr. Cane has also devoted considerable effort to studying the impacts of El Niño, especially its impact on agriculture. His 1994 paper showing that El Niño has a strong effect on the maize crop in Zimbabwe has helped that country to factor climate variability into its agricultural planning decisions. In addition, Dr. Cane's pioneering work in short-term climatic forecasting was instrumental to the founding in 1996 of the IRI for Climate Prediction at the Lamont Observatory. The main purpose of the IRI is to create a global research and forecasting network "to accelerate advances in predicting short term climate and speed the practical application of that skill around the world" (Stevens, 1996) A more recent research interest of Cane's has been the reconstruction and study of past climates, with the goal of learning how Earth's climate has varied historically, to gain a better understanding at the influence humans may have on present day climate.

Cane was born and raised in Brooklyn, New York, the son of an engineer and a school attendance officer. While at Midwood High School, he developed an interest in math, which he pursued further at Harvard University, majoring first in pure mathematics, then switching to applied mathematics. He obtained both his BA (1965) and MS (1968) in applied mathematics from Harvard. While an undergraduate, he took many social science courses, and became a political activist. Just after graduation, in the summer of 1965, he went to Alabama and joined the civil rights crusade. Back at Harvard in pursuit of his masters' degree, he did work in computer science, and on obtaining his degree, he went to work for the NASA Goddard Institute for Space Studies in New York as a computer programmer. He then married Barbara Haak, a lawyer; they had two children.

After a short stint teaching math at New England College in Henniker, New Hampshire, Cane went to MIT to study under the famous meteorologist, Jule D. Charney. There, Cane worked on a computer model of an equatorial ocean current, thus laying the groundwork for his future modeling of the El Niño phenomenon. He obtained his PhD from MIT in 1976, returning there as a faculty member in 1979. He then began working with his graduate student, Stephen E. Zebiak, to reproduce, via computer model, the oscillations in equatorial ocean temperature characteristic of El Niño. In the mid-1980s, the two researchers moved to the Lamont Observatory, continuing to refine and test their predictive model. The Cane–Zebiak model has successfully predicted most fluctuations of El Niño's cycle since 1986–1987 (Cane 2000, personal communication; Stevens, 1996).

Cane has written close to 150 papers in oceanography and climatology. In 1992, he received the Sverdrup Gold Medal awarded by the American Meteorological Society. Hobbies have included squash, tennis, and kayaking.

JAMES J. O'BRIEN: August 10, 1935–

Dr. James J. O'Brien is a professor of meteorology and oceanography at Florida State University (FSU) in Tallahassee, where he has been a faculty member since 1969. He is also director of the Center for Ocean-Atmospheric Prediction Studies (COAPS) at FSU. Internationally recognized for his expertise in the computer modeling and predicting of El Niño events, as well as their impacts, O'Brien is a prolific researcher. Together with his students and collaborators, he has published more than 200 articles on topics ranging from coastal upwelling to the measurement of oceanic wind fields from satellite borne scatterometers to the Pacific Ocean's effect on atmospheric carbon dioxide, in addition to his many papers on El Niño. Since 1990, O'Brien and his students have discovered that El Niño reduces the number of Atlantic hurricanes as well as the number of Tornado Alley tornadoes. They have also done extensive studies on how El Niño and La Niña affect rainfall patterns, the occurrence of forest fires, and farm crop productivity in the United States.

O'Brien was born in the Bronx, New York, the first child of Irish immigrants, who met in 1927 during their passage to the United States. The family moved to New Jersey shortly after his father was transferred to the Bell Telephone Laboratories in Murray Hill, New Jersey in 1940. James graduated from Summit High School in 1953, and enrolled as a chemistry major at The Mens College, Rutgers University, in New Brunswick, New Jersey. In June 1957, he graduated with a BS as well as a second lieutenant commission in the Air Force, having participated in the Air Force ROTC while in college. Because the Korean War had just ended, and he was not needed for immediate military duty, he joined E.I. duPont in Niagara Falls, New York. When the Air Force reclaimed him a year later, they assigned him to be a meteorologist and sent him for a year's meteorological training at the University of Texas in Austin, where according to O'Brien's own account, "he found his love in science."

Having made the transition from chemistry to meteorology, O'Brien proceeded on to graduate school at Texas A&M at the age of 29, and received both master's and PhD degrees in 3½ years. Beginning with this work, O'Brien began to delve into the interaction between atmosphere and ocean, a theme that would follow through much of the research that followed. As he described his doctoral work, he "explained how a hurricane pumps up cold ocean waters, which kill the storm if the storm does not move." After his graduate studies were completed, he worked at the NCAR in Boulder, Colorado for 3 years, before accepting his faculty position at FSU, where he attained full professorship by the age of 39.

O'Brien has received numerous honors and awards. Among them, he was elected to the Russian Academy of Natural Sciences, and the Norwegian Academy of Science and Letters. In 1999 he received FSU's Robert O. Lawton Distinguished Professor Award, the highest honor FSU bestows on faculty. He has three living children and two grandchildren. Fishing is his main hobby.

EUGENE M. RASMUSSEN

Eugene M. Rasmussen is a senior research scientist in the Department of Meteorology at the University of Maryland, and a past president

(1998) of the American Meteorological Society (AMS). In 1982, he published, with co-author T. H. Carpenter, a landmark paper on ENSO entitled, "Variations in tropical sea surface temperature and surface wind fields associated with the Southern Oscillation/El Niño." Common use of the term *ENSO* linking El Niño with the Southern Oscillation followed the publication of this paper. In support of the study used in this paper, the National Climatic Center compiled a new set of sea surface temperatures (SSTs) and marine surface winds. This dense and lengthy data set included Pacific marine observations from many shipping nations and agencies over the period from 1854 through 1976, and permitted comparison of SST variation with respect to the SOI. Rasmussen also devised, in the early 1980s, the widely used NIÑO 1, 2, 3, and 4 regions (Allan, 1996).

Rasmussen received his undergraduate education in engineering, but became an Air Force meteorologist soon after graduation. On completion of his military service in 1955, he joined the U.S. Weather Bureau, first as a hydrologist, then as a state forecaster, attending graduate school at the same time. In 1963, he obtained his MS in engineering mechanics from St. Louis University. In 1964, Rasmussen began his research career at the Geophysical Fluid Dynamics Laboratory of the Environmental Science Services Administration (ESSA). ESSA was the successor to the Weather Bureau, and predecessor to the current NOAA. He received his PhD in meteorology from MIT, and in 1972, he joined the Center for Experiment Design and Data Analysis as the chief of the research division.

In 1979, he became the chief of the Diagnostics Branch of the NOAA Climate Analysis Center (CAC). Rasmussen joined the University of Maryland in 1986, and continues his research there. He has been continuously active in the AMS, served as editor of three AMS journals, and received some of the society's most prestigious awards. In addition, he has served on many advisory panels, committees, boards, and commissions.

CHESTER F. ROPELEWSKI: May 11, 1942–

Chester F. Ropelewski is the director of Climate Monitoring and Dissemination at the IRI for Climate Prediction within the Lamont-Doherty Earth Observatory of Columbia University. He has published more than 50 scientific papers, including several on the influence of ENSO on patterns of rainfall and temperature. In addition, he has written many reports on climate, climate variability, and climate change issues, and he currently chairs the AMS Climate Variations Committee. In 1983, he co-authored with M. S. Halpert a paper entitled, "North American Precipitation Patterns Associated with the El Niño/Southern Oscillation (ENSO)," which is frequently cited as a pioneering work. The paper presents the seasonal response to El Niño of temperature and precipitation patterns across various regions of North America.

Ropelewski grew up in western Massachusetts, with an early scientific interest in geology. However, when he went to college, he chose physics as his major. After graduation, he spent 5 years in the military, where he received his introduction to atmospheric science as a military meteorologist. Having discovered that he enjoyed this field of science, he returned to school as a graduate student in meteorology at Penn State University. He later continued his career as a forecaster with the National Weather Service (NWS), and also worked in air pollution meteorology, applying meteorological analysis to atmospheric transport and dispersion. Following this, he joined the NWS Climate Prediction Center (CPC) as a research meteorologist, and became branch chief of the Analysis Branch of the CPC in 1990. In 1997, he left that post to assume his current position at the Lamont-Doherty Earth Observatory.

Ropelewski was greatly influenced by the work of Rasmussen regarding the possible relationship between tropical ocean temperatures, and the weather and climate at mid-latitudes, and he has devoted much of his career to investigating this linkage. Another area of interest has been the application of related theories and discoveries to the prediction of seasonal climate variability. Furthermore, his early experience in analysis and forecasting has given him a heightened interest in making our newly found ability to predict seasonal climate fluctuations more useful to those who might benefit from such predictions.

SIR GILBERT WALKER: June 14, 1868– November 4, 1958

As the man who discovered the first piece in the El Niño puzzle, Sir Gilbert Walker occupies a pre-eminent place in the history of global climate research. A British meteorologist and applied mathematician,

Figure 12.2.   Sir Gilbert Thomas Walker
Courtesy of Eugene Rasmussen, University of Maryland.

he was the first to observe the oscillations in pressure, temperature, and rainfall that occur from one side of the tropical Pacific to the other every few years, and it was Walker who named this phenomenon the Southern Oscillation. However, Walker was not seeking to understand the workings of El Niño, but rather to understand the mechanisms that govern the strength of the Indian monsoon, with the ultimate goal of being able to predict its variation from year to year.

Walker's interest in the Indian monsoon sprang from his appointment in 1904 as the director-general of observatories in India, under which title he administered the Indian state meteorological service. Because a lack of monsoonal rainfall had recently resulted in the great Indian famine of 1899–1900, he was keenly aware of the monsoon's effect on the country's agriculture. In his quest to find a way to predict the variation in the monsoon from year to year, he examined extensive sets of meteorological data, applying rigorous statistical tests in his analysis of this data. Although he was never able to achieve his goal of forecasting the strength of the Indian monsoon, his discovery of the Southern Oscillation was significant in laying the groundwork for the El Niño research that was to come later.

Walker had arrived in India well prepared to conduct groundbreaking meteorological research, equipped with a background firmly rooted in mathematics and science. Having attended Trinity College, Cambridge on a mathematical scholarship, he became a lecturer there in 1895. At that time, his main interest was electromagnetism, and he published a series of papers on that subject. Walker also had a lifelong interest in both the physics of projectiles and the physics of flight. He earned the nickname "Boomerang Walker" because of his studies of such ancient projectiles as the boomerang and the stone-age celt. His interest in flight was inspired by the soaring and gliding of the Himalayan' birds, which he observed in India. He first studied the natural flight of birds, then extended his work to human soaring and gliding. In addition, Walker was interested in the physics of music production, and was responsible for improvements in the design of the flute.

After retiring from his post in India in 1924, Walker became a professor of meteorology at the Imperial College of Science and Technology in London. There he continued to investigate connections between the weather in different parts of the world, as well as other topics such as cloud formation. He retired from that position in 1934, but remained scientifically active well into his 80s.

Throughout his life, he received many honors, chief among them being knighted in 1924. He was president of the Royal Meteorological Society from 1926 to 1928, and editor of the *Quarterly Journal of the Royal Meteorological Society* from 1935 to 1941.

Walker married May Constance in 1908; they had one son and one daughter. According to one acquaintance, Sir Gilbert Walker "ever remained modest, kindly, liberal minded, wide of interest, and a very perfect gentleman."

HURD CURTISS WILLETT: January 1, 1903– March 29, 1992

Because of the success of his long-range forecasts, Hurd Willett was "perhaps the most widely quoted and widely known meteorologist in the history of science," according to an article by Clifford H. Nielsen in an issue of *Weatherwise*. A long-time professor of meteorology at MIT, Willett's pursuit of improved forecasting techniques spanned not only his lifetime, but also the 20th century.

Willett was born in Providence, Rhode Island to parents Mabel Hurd and Allen Willett, both of whom had earned PhD's from Columbia University. He was raised and home schooled at the family farm in a small town north of Pittsburgh, Pennsylvania. There he developed his love of weather, making daily entries in his weather diary to document some particularly cold winters the family experienced.

At the age of 17, Hurd made a trip to visit the U.S. Weather Bureau in Washington, DC, and on the recommendation of Walter J. Humphreys, whom he met there, he enrolled at Princeton to study math and physics. After graduation, he went back to Washington, DC to study under Humphreys at George Washington University, while at the same time being employed as a weather observer with the Weather Bureau. There he impressed Carl-Gustav Rossby (a very famous meteorologist who did groundbreaking work in the study of atmospheric waves) to the extent that Rossby obtained for him the first Guggenheim Fellowship for foreign study. In the spring of 1927, Willett used this fellowship to travel to the Bergen Institute in Norway to learn about the polar front theory that had recently been developed there.

In 1928, Willett and Rossby both went to Cambridge, Massachusetts to start a new program in meteorology at MIT, where Willett ran the synoptic meteorology laboratory, and also began his research on global circulation patterns, applying what he had learned about polar front theory in Norway. In collaboration with Rossby and research assistants Jerome Namias and Victor Starr (both of whom subsequently became very well-known and successful meteorologists), Willett developed a method of making 5-day forecasts. This method was adopted by the Weather Bureau in 1939 when Namias left MIT to become head of the Weather Bureau's long-range forecasting section. In recognition of this work, Willett received the first Distinguished Scientific Achievement Award from the AMS in 1951. During World War II, Doc Willett helped train military forecasters, consulting at army bases throughout western Europe.

After the war, Willett turned his research efforts to another area that would dominate the remainder of his career, that of investigating the relationship between climate fluctuations and changes in solar output during sunspot cycles. His hope was that his hypothesis of such a correlation could bring some order and predictability to the chaotic nature of weather occurrences, including the irregular occurrences of large-scale climate fluctuations such as the ENSO. By the 1970s, he was working with statistician and former student John Prohaska to develop advanced correlations between the atmospheric circulation and solar cycles, which he then used to create his long-range and seasonal forecasts.

Willett continued at Cambridge for 59 years, until retiring from active forecasting and research in 1988 at the age of 85. Although he was never able to establish a physical link between solar variations and changing weather patterns, he was able to make successful long-range predictions using the correlations he developed. According to Harold W. Bernard, who interviewed him for a 1980 article in the *National Weather Digest*, "Dr. Willett is one of the few credible long range forecasters who will talk about weather trends beyond one season."

KLAUS WYRTKI: February 7, 1925–

Dr. Klaus Wyrtki, professor emeritus of the University of Hawaii Department of Oceanography, is known as a pioneer in the study of large-scale oceanographic changes, as well as interactions between the oceans and the atmosphere. According to the AMS on the occasion of his receipt of that organization's Sverdrup Gold Medal, "His classic paper entitled 'El Nino—The Dynamic Response of the Equatorial Pacific Ocean to Atmospheric Forcing,' is the most widely cited paper in the American Meteorological Society's *Journal of Physical Oceanography.*"

In keeping with his love of data collection and interpretation, Wyrtki devised an array of 33 Indo-Pacific island observation sites for measuring temperature, wind, and sea-level heights. With this array, he was able to monitor and study oceanic phenomena to a degree not possible before.

Among other notable contributions to the study of the ENSO, he introduced in 1975 the concept of "westerly bursts," which he hypothesized to eventually induce the onset of an El Niño event. This concept represented a modification to the earlier model developed by Bjerknes, and was developed on the strength of his extensive oceanographic data set. Just prior to the 1972 El Niño event, his data showed a build up in sea level of as much as a foot and a half along

Figure 12.3.   Klaus Wyrtki
Courtesy of Klaus Wyrtki.

the coasts of southeast Asia, together with a simultaneous drop in sea level along the western coast of South America. Wyrtki proposed that the initial increase in sea-level height over the western Pacific was produced by stronger than usual westward blowing trade winds (westerly bursts), which pushed warm surface water westward. He observed that when this thick bulge of warm water oscillated eastward, amplifying the Equatorial Countercurrent, it would eventually increase the height of the ocean surface in the eastern Pacific. In addition, the depth of the warm water in the eastern Pacific would increase, thus initiating an El Niño event.

Wyrtki was born in Tarnowitz, Germany, and received all his formal education in Germany. In 1945, just after the end of World War II, he enrolled at the University of Marburg to study physics and mathematics, subjects he chose after discovering that ship building (his first choice) was no longer being taught in Germany. In 1948, he began his advanced studies at the University of Kiel, from which he received his doctor of natural science in oceanography, physics, and mathematics in 1950. He went on to do some post doctoral research at the University of Kiel until 1954, when he became the first director of the Institute of Marine Research in Djakarta, Indonesia. There he began to study the oceanography of equatorial regions, work he continued during his 3 years (1958–1961) as senior research officer at the Commonwealth Scientific and Industrial Research Organization in Sydney, Australia.

From 1961 to 1964 he worked as a research oceanographer at the Scripps Institution of Oceanography in La Jolla, California; he then joined the faculty at the University of Hawaii, where he remained until his retirement in 1993.

Dr. Wyrtki has won numerous awards and served on the scientific panels of several international organizations. He was an editor of the *Journal of Physical Oceanography* from 1971 to 1979. He is the author of more than 100 research publications. The Wyrtki Center at the University of Hawaii at Mano was named to honor him. He is married, with one son and one daughter, and became a naturalized U.S. citizen in 1977.

STEPHEN E. ZEBIAK: July 24, 1956–

Dr. Stephen E. Zebiak is currently director of the modeling and prediction research branch of the IRI for Climate Prediction. This institute, which Zebiak helped to found, is part of the Lamont-Doherty Earth Observatory in Palisades, New York, which in turn, is associated with Columbia University. Dr. Zebiak's most notable contribution to date is in the area of computer modeling of the ENSO phenomenon. In the mid-to late 1980s, he co-authored (with Mark Cane) the first dynamic computer model successful in predicting an El Niño event (Cane, Zebiak, & Dolan 1986). The model produced the "first equatorial Pacific sea surface temperature forecasts, and is still the most widely used intermediate model" (Allen et al., 1996). It is used for routine diagnosis and prediction of ENSO fluctuations, and because of its simplicity, "can be run over long periods of time and used for testing various ideas" (Perigaud & Dewitte 1996).

Zebiak was born in a small town in Maine, one of four children. His father was a chemical engineer working in the paper industry there. Influenced and encouraged by his high school mathematics and physics teachers, he went on to study applied mathematics at the MIT. While at MIT, he took an elective course in meteorology, taught by Professor Fred Sanders, which prompted him to see that he could apply his math background to a field he had always found interesting. (While in junior high, he had submitted a winning science fair project relating to weather)

Zebiak then went on to receive his master's degree in applied mathematics at Rensselaer Polytechnic Institute, and returned to MIT for his doctoral work. His advisor there, Dr. Mark Cane, got him interested in the subject of El Niño, and in 1984, Zebiak obtained his PhD in meteorology from MIT. Since that time, he has worked in the area of ocean-atmosphere interaction and climate variability. Hobbies include playing the piano.

Chapter 13

Directory of Organizations

Following is a *basic* list of organizations that are involved with ENSO. Entering the name of any of the organizations into a Web search engine will get one to the latest Web sites for these organizations. While there, one might check for links and partners on their sites to find additional groups and organizations not on this list.

It should be noted that many universities with meteorology or climate programs conduct research or maintain information on their Web sites about ENSO and their ENSO research.

ENVIRONMENTAL AND SOCIETAL IMPACTS GROUP

Environmental and Societal Impacts Group
National Center for Atmospheric Research
PO Box 3000
Boulder, CO 80307
www.esig.ucar.edu

The Environmental and Societal Impacts Group is by NOAA's Office of Global programs, which focuses on issues related to ENSO and its effects on ecosystems and societies. The group publishes the results of their assessments in the popular journals and in their newsletter, the *ENSO SIGNAL.*

CLIMATE PREDICTION CENTER (CPC)

Climate Prediction Center (CPC)
5200 Auth Road, Room 800
Camp Springs, MD 20746
www.cdc.noaa.gov

In the 1980s the NWS established the CPC, known at the time as the Climate Analysis Center, located in Camp Springs, Maryland. CPC is best know for its U.S. climate based on El Niño and La Niña conditions in the Tropical Pacific. CPC serves the public by assessing and forecasting the impacts of short-term climate variability, emphasizing enhanced risks of weather-related extreme events, for use in mitigating losses and maximizing economic gains.

The CPC's products are operational predictions of variability, real-time monitoring of climate and the required databases, and assessments of the origins of major climate anomalies. The products cover scales from a week to seasons, extending into the future as far as technically feasible, and cover the land, the ocean, and the atmosphere, extending the stratosphere.

These climate services are available for users in government, the public and private industry, both in this country and abroad. Applications include mitigation of weather-related natural disasters and uses for social and economic good in agriculture, energy, transportation, water resources, and health. Continual product improvements are supported through diagnostic research, increasing use of models, and interactions with user groups.

Climate Diagnostics Center (CDC)
Climate Diagnostics Center
NOAA/ERL/CDC
MC: R/E/CD
325 Broadway
Boulder, CO 80303
Fax: (303) 497–7013
www.cdc.noaa.gov

The mission of the CDC is to identify the nature and causes of climate variations on time scales ranging from a month to centuries.

The goal of this work is to develop the ability to predict important climate variations on these time scales. Short-term climate variations of interest include major droughts and floods over the continental

United States, and the global anomalies associated with ENSO. These events attract great public interest, and often have enormous social and economic consequences. On longer time scales, basic research goals include identifying the causes for decadal to centennial climate variations, and separating natural variability from anthropogenically induced climate changes in order to provide an improved scientific basis for public planning and policy decisions. CDC has made considerable progress toward these goals through a coordinated program of diagnostic and modeling studies aimed at significantly advancing understanding and predictions of climate variability.

Climate Research Division (CRD)

Climate Research Division
Scripps Institution of Oceanography
University of California, San Diego
9500 Gilman Drive, Dept. 0224
La Jolla, CA 92093–0224
meteora.ucsd.edu/crd.html

Scripps Institution of Oceanography is playing a leading role in pioneering the interdisciplinary study of the earth as a unified system. In the CRD, scientists study a broad range of phenomena. These span time scales from a few weeks to several decades. Research themes include predicting the natural variability of climate and understanding the consequences of man-made increases in the greenhouse effect. Climate change caused by human actions is the paradigm that illustrates why traditional disciplinary barriers in the earth sciences are rapidly weakening. In the climate system, the atmosphere, the seas, the land surface, and the world of living things are tightly coupled. To understand these interactions, a variety of expertise must be brought to bear through a team approach to research.

Current research projects include the development of coupled global ocean and atmosphere models, assessing the role of cloud-radiation feedbacks in climate change, and modeling and predicting seasonal climate variability. CRD research combines the analysis of large observational data sets, the development of comprehensive numerical models of the climate system, and the exploitation of satellite remote sensing capabilities for monitoring the entire planet. CRD researchers collaborate closely with other scientists at Scripps and elsewhere.

Studies focus on a wide range of regional and global climate phenomena, including ENSO, the Indian monsoon, the Pacific Intertropical Convergence Zone, the California Current system, and precipitation and water supply in the United States.

CRD scientists have stressed research on the regional and transient implications of global change for climate, emphasizing those aspects of climate that are potentially predictable. In recent work, CRD scientists and their collaborators have developed advanced coupled ocean-atmosphere models for ENSO prediction. This research is critical to global change objectives, because there are strong indications that climate changes such as greenhouse warming may have profound effects on ENSO phenomena.

European Centre for Medium Range Weather Forecasts (ECMWF)
ECMWF
Shinfield Park
Reading RG2 9AX
UNITED KINGDOM
www.ecmwf.int

The ECMWF is an international organization supported by 18 European States: Belgium, Denmark, Federal Republic of Germany, Spain, France, Greece, Ireland, Italy, Yugoslavia (inactive since June 5, 1992), the Netherlands, Norway, Austria, Portugal, Switzerland, Finland, Sweden, Turkey, United Kingdom.

The ECMWF has concluded cooperation agreements with Croatia, Iceland, Hungary, and Slovenia, and has working arrangements with the World Meteorological Organisation (WMO), the European Organisation for the Exploitation of Meteorological Satellites (EUMETSAT), and the African Centre for Meteorological Applications for Development (ACMAD).

Originally a European Cooperation in Science and Technology project, the ECMWF was established in 1973 by a convention. The first real-time medium-range forecasts were made in June 1979. The ECMWF has been producing operational medium-range weather forecasts since August 1, 1979.

NOAA Office of Global Programs (OGP)
NOAA Office of Global Programs
1100 Wayne Avenue, Suite 1210
Silver Spring, MD 20910
301–427–2089
Fax: 301–427–2073
www.ogp.noaa.gov

The OGP leads the NOAA Climate and Global Change (C&GC) Program. OGP assists NOAA by sponsoring focused scientific research aimed at understanding climate variability and its predictability. Through studies in these areas, researchers coordinate activities jointly contribute to improved predictions and assessments of climate variability over a continuum of time scales from season to season, year to year, and over the course of a decade and beyond.

Center for Ocean-Land-Atmosphere Studies (COLA)
Center for Ocean-Land-Atmosphere Studies
4041 Powder Mill Road, Suite 302
Calverton, MD 20705–3106 USA
(301) 902–1254
Fax: (301) 595–9793
grads.iges.org/home.html

The center is a group of uniquely qualified scientists dedicated to understanding the problem of seasonal to interannual and decadal climate fluctuations with special emphasis on the role of interactions between earth's oceans, atmosphere, and land surface.

The primary goal of COLA is to foster interdisciplinary research and to increase understanding of the physical processes in the atmosphere, at the land surface, in the oceans, and the interactions among these components. It is recognized that the interactions among atmospheric, oceanic, and land surface processes are perhaps the most important determinants of the interannual variability of the present climate that affects the global and regional habitability of the planet Earth. A better understanding of interactions among these processes is essential to enable researchers to distinguish between the natural variability of the coupled system and changes caused by external forcing and human activities.

An important objective of the center is to study the contributions of internal dynamic processes and the slowly varying boundary conditions at the earth's surface in determining the variability and predictability of climate on the short term, and to explore the feasibility of dynamic prediction of seasonal to interannual variations. The focus of the center is to understand the mechanisms underlying climate variations on time scales of months to years and longer and to determine the limits of predictability of climate at these time scales. The center utilizes comprehensive physical models of atmospheric, oceanic, and land surface processes to carry out sensitivity and predictability studies.

The center has a strong capability for theoretical work on atmospheric dynamics, diagnosis of observed climate data, and computer modeling of the global ocean–atmosphere–land system.

A global general circulation model (GCM) of the atmosphere (AGCM) is used to study the sensitivity to prescribed changes in the sea surface temperature, snow cover, sea ice, and soil moisture at the earth's surface. A global biosphere model has been developed and coupled with the atmospheric model. Global and basin versions of an ocean general circulation model (OGCM) are being used for studies of data assimilation and predictability of the oceans. These studies serve the dual purpose of increasing our understanding of air–sea and air–land interactions, and defining the accuracy and resolution requirements for space observing systems designed to measure these boundary conditions.

The dynamics of tropical and mid-latitude droughts, and the climatic effects of deforestation and desertification are topics of interest at COLA. The center also studies the dynamics of the Asiatic monsoon and its interactions with the planetary scale circulations such as those associated with ENSO. Predictability studies using a global GCM have suggested that, for seasonal to interannual time scales, the tropical atmosphere is potentially more predictable than the mid-latitudes because changes in the boundary conditions produce large changes in the tropics. Since the influence of tropical anomalies can propagate to the mid-latitudes, this also provides some hope for predictability in the extratropics. The mid-latitude summer droughts are also potentially predictable because the variations due to internal dynamics are relatively weaker during the summer season.

COLA provides a unique opportunity for scientists engaged in atmospheric, oceanic, and land surface modeling to work together in a scholarly environment toward a common objective of developing a coupled ocean–atmosphere–land model, while pursuing independent research in each of the individual areas. The current emphasis of the center is on the physical problems of sea–air–land interaction rather than the biological or chemical aspects. Although most academic institutions specialize in the individual disciplines of meteorology, hydrology, and oceanography, it is desirable to create a nucleus of scientists with expertise in each of these fields who can work together in interdisciplinary research on ocean–atmosphere–land interactions.

International Research Institute (IRI)
International Research Institute

Director at LDEO
Columbia University
204 Oceanography Building
Route 9W
Palisades, NY 10964
iri.ldeo.columbia.edu

The IRI for climate prediction was established through a cooperative agreement NOAA/OGP and Columbia University's Lamont-Doherty Earth Observatory (LDEO). The mission of the IRI is to continually assess and develop seasonal to climate forecasts, and to foster the application of such climate forecasts to the benefit of societies. The IRI will address all aspects of end-to-end prediction, including model and forecast system development, experimental prediction, climate monitoring and dissemination, applications research, and training, in coordination and collaboration with the international climate research and applications community. The IRI will eventually assume multinational governance, and will coordinate with meteorological and hydrological services and other agencies in the delivery of forecast products and the establishment of climate applications activities on a worldwide basis.

Center for Atmosphere Prediction Studies

Center for Atmosphere Prediction Studies
Florida State University
COAPS Suite 200, Johnson Building
Tallahassee, FL 32306–2840
www.coaps.fsu.edu

COAPS performs research in air–sea interaction, ocean and air–sea modeling, climate prediction on scales of months to decades, statistical studies and predictions of social economic consequences of the ocean–atmospheric variations. COAPS studies the regional and national effects that ENSO events have on weather patterns and weather phenomena such as hurricanes, tornadoes and winter snows.

Pacific ENSO Applications Center (PEAC)

Pacific ENSO Applications Center
University of Hawaii at Manoa
2525 Correa Road, HIG Room 350
Honolulu, HI 96822
(808) 956-2324
Fax: (808) 956-2877
lumahai.soest.hawaii.edu/Enso/

PEAC was established in 1994 as a multi-institutional partnership to conduct research and produce information products on climate variability related to ENSO climate cycle in the U.S.-affiliated Pacific Islands. The focus is on ENSO's historical impacts and the latest long-term forecasts in support of planning and management in such climate sensitive sectors as water resource management, fisheries, agriculture, civil defense, public utilities, coastal-zone management, and other economic and environmental sectors of importance to the communities.

Chronology

The current understanding of El Niño did not come easy. In fact, for many decades, there were two parallel research efforts, one focusing on the atmosphere and the pressure oscillations of the Southern Oscillation, and the other on the oceans and the periodic warming known as El Niño. It was not until the 1960s that researchers finally understood that the atmosphere and ocean were acting in concert and that the El Niño and Southern Oscillation were part of the same phenomenon.

Year	El Niño or Southern Oscillation	Milestone
15th century	El Niño	Potato farmers in the mountains of Peru and Bolivia were able to predict El Niño and the associated cycles of drought and flood by observing the night sky in the months before the growing season.
16th century	El Niño	The term *El Niño* is born. It was first used by 16th-century Peruvian fishermen to refer to a warm southward moving ocean current typically appeared each year around Christmas time.
1877	Southern Oscillation	The worst famine in India's history, caused by the failure of the monsoon in a strong El Niño, triggered the first intensive research work. An observatory is founded in India to explore the cause and whether the famines could be prevented

Year	El Niño or Southern Oscillation	Milestone
1897	Southern Oscillation	H.H. Hildebrandsson is the first to postulate the existence of a large-scale atmospheric oscillation in the tropical Pacific based on an examination of 10 years of surface pressure data.
1902	Southern Oscillation	Norman and W.J.S. Lockyer confirm the apparent flip-flop pressure changes using more extensive data.
1904	Southern Oscillation	Sir Gilbert Walker, a British mathematician entered the British Colonial Service as the director general of the observatory.
1920s	Southern Oscillation	Sir Gilbert Walker and E. W. Bliss, while studying the Indian monsoon, further documented this oscillation using a full 40 years of station data. Walker named it the Southern Oscillation. He devised an index that later became known as the Southern Oscillation Index (SOI). He also noted that there was a high correlation between the index in the June through August period to the following December through February period. This led Walker to correctly observe that the index would be useful in long-range forecasting.
1921	El Niño	Brooks and Braby presented evidence of a relationship between El Niño warm events and rainfall, winds, and temperatures.
1925	El Niño	A strong oceanic El Niño warm event in 1925 devastated the Peruvian populations and ecosystems. Peru was then motivated to start gathering rainfall and other weather information to go along with the sea temperature data.
1933	Southern Oscillation	J.B. Leighly put the two atmospheric pieces together. He related the changes in rainfall to changes in the trade winds, which in turn he related to the pressure changes of the Southern Oscillation.
1952	Southern Oscillation	Willet and Bodurtha devised a much simpler formula for the SOI focusing on two locations (Darwin, Australia and the South Pacific island of Tahiti). This is used today.
1956	Southern Oscillation	Schell looked at local sea temperatures and their effect on the South Pacific high pressure center, the strength and position of which determines the strength of the trade winds. He proposed that the

Year	El Niño or Southern Oscillation	Milestone
1960s	Both	"oscillatory" nature of the Southern Oscillation could be the result of a cyclic feedback process. Jacob Bjerknes, working with data gathered during the 1957 International Geophysical Year, noted that the significant oceanic warming of the strong 1957–1958 El Niño coincided with a strong negative phase of the Southern Oscillation. Thus, ENSO was born.
1960s	Both	Bjerknes proposed the link between the ocean and atmosphere in the form of a direct circulation, which he named the Walker Circulation.
1970s	Both	Klaus Wyrtki, an oceanographer at the University of Hawaii discovered the interrelationship between changes in the Pacific Equatorial countercurrent and ENSO
1982–1983	Both	The Great El Niño of 1982–1983 moved El Niño from the technical journals to the popular media and research institutions. Worldwide catastrophes accompanied the great warming that year.
1980s	Both	Ropelewski and Halpert of the U.S. National Weather Service's Climate Analysis Center found significant relationships between precipitation and the two phases of the ENSO in many diverse areas of the globe. They discovered that the effects were very nearly the opposite in one phase from the other.
1986	Both	The first successful forecast El Niño event was made by Cane and Zebiak, of Lamont Observatory.
1997	Both	The first bold forecast of a major El Niño is made 6 months in advance of the Super El Niño of 1997–1998 by the Climate Prediction Center and other forecasters.
1999	El Niño	Researchers at Jet Propulsion Lab and the University of Washington propose that a larger scale oscillation exists in the Pacific which changes on a multi-decadal scale. ENSO is a smaller time scale flip-flop within the larger scale oscillation which they called the Pacific Decadal Oscillation (PDO).

Glossary

Atmospheric pressure is the weight of a column of air from the surface to the top of the atmosphere. **Wind**, which is air in motion, is driven by differences in pressure along the surface.

A **Cool Event** (or alternately **Cool Phase of ENSO** or alternately **ENSO Cool Event**) is a La Niña or an anomalous cooling of the waters in the eastern and central Tropical Pacific. Cooling of the waters there is often accompanied by an accumulation of warm water in the western Tropical Pacific.

Convection is a process of heat transfer involving vertically moving air currents. It is most favored in unstable air where low level warm air becomes buoyant and rises. Cooler, heavier air from above sinks to replace it. Convection results in clouds and showery type precipitation. In La Niñas, the convection is confined mainly to the western Pacific. In El Niños, the convection in the Tropical Pacific shifts eastward, which has an effect on atmospheric jet streams and storm tracks.

Coral Bleaching occurs when water temperatures rise. Coral reefs are unique and very rich ecosystems found in the tropical oceans that supports a wide variety of marine life (animal and plant). Coral thrives at temperatures below 28C (82F). When water temperatures rise above 28C, however, a process known as coral bleaching occurs, during which the coral expels the algae (microscopic organisms called Zooxanthallae) that thrive in the coral and are necessary for its survival. The coral and the algae have a special symbiotic relationship in which the algae supply oxygen and some organic compounds to their coral hosts. When the algae is expelled, the coral polyps can lose their pigmentation and appear transparent on the animal's white skeleton. If the warm temperatures persist, other marine life that thrive in the coral are also affected and ultimately the quality of the fishing and recreation in the region.

The **Coriolis Force** is an apparent deflecting force for objects in motion relative to the earth that is caused by the earth's rotation. Air in the atmosphere and water in the oceans that is moving relative to a point on the earth will be apparently deflected to the right in the northern hemisphere and to the left in the southern hemisphere. The Coriolis force is zero at the equator. The Coriolis Force was first described mathematically by Gustave Gaspard Corilis, a French scientist in 1835.

El Niño was a term originally used by 19th-century fisherman to describe a southward moving current of warmer water off the coast of Peru and Ecuador that occurred each year around or shortly after Christmas. The term *El Niño* means "the Infant Boy" referring to "the Christ Child." In some years, the warming is more extensive and can last 1 or 2 years. Today, El Niño is used to describe these large-scale, long-lasting "warm" events. El Niño is characterized by large-scale weakening of the trade winds and warming of the surface layers in the eastern and central equatorial Pacific waters. These events occur irregularly every 2 to 7 years (averaging every 3 to 4 years).

ENSO is an acronym that stands for El Niño Southern Oscillation. The term is used to describe the full range of the Southern Oscillation including both the warm (El Niño) and cold (La Niña) events. Technically, it includes both the oceanic (warm or cool water) component and the atmospheric (Southern Oscillation) component of the phenomena. Sometimes the term has been incorrectly used to refer only to the warm events or El Niños.

Hadley Cell is the three-dimensional circulation of air in the tropics and subtropics. Air rises in the Intertropical Convergence Zone (ITZ) and then moves poleward at high altitudes. It then converges in the subtropics with air moving equatorward from higher latitudes and sinks in the Subtropical High belt. When it reaches low levels, the air then moves equatorward as the trade winds only to converge and rise again at the ITCZ. This circulation helps produce the clouds and rainfall and low pressure of the ITCZ and the sunshine, high temperatures, and high pressures of the subtropical high zone.

Intertropical Convergence Zone (ITCZ) is the belt of clouds and showers that oscillates seasonally around the equator. It results from a convergence of the trade winds from both hemispheres. The relative strength and level of activity associated with the ITCZ varies considerably in the El Niño and La Niña years. In El Niños, the ITCZ is anomalously active in the eastern and central Pacific but anomalously weak in the west. In La Niñas, the opposite is observed.

The **Jet Stream** is that high-speed ribbon of air that generally blows from west to east in middle latitudes. In winter, the winds at the jet stream level, from 5 to 10 miles up, can blow at 200 mph, equaling the strongest winds ever recorded at the surface in hurricanes! These winds are strongest in the zone of greatest temperature contrasts. That is also where the storms tend to track.

A **Kelvin Wave** is a very long (1,000s of kilometers in length) wave that moves east across the Equatorial Pacific. Satellites today tell us these waves cross the Pacific in 2 to 3 months. They affect the thickness of the warm water, the depth of the thermocline (lowering it by 30 meters or more) and can cause large changes in sea level (tens of centimeters). Kelvin waves can lead to El Nino by transporting warm western tropical Pacific water to the eastern tropical Pacific. When Kelvin waves reach South America they can generate long-lived Rossby waves that travel west and ultimately cause the demise of an El Nino event.

La Niña (in Spanish meaning "the Infant Girl") is in many respects the opposite of El Niño. It has also been called El Viejo ("the Old One") and the Anti-El Niño, but this term has fallen into disfavor as it would imply an "anti-Christ." This cool relative of El Niño is characterized by a large-scale strengthening of the trade winds and a cooling of the water in the eastern and central tropical Pacific due to upwelling of cool water from beneath. Like El Niños, La Niñas often begin in the summer of the southern hemisphere (December to February) and may last a year or two. Like the El Niño, the La Niña has occurred irregularly between 2 and 7 years, averaging once every 4 or 5 years.

Madden-Julian Oscillation is a quasiperiodic oscillation of the near equatorial troposphere, most noticeable with the wind near the surface and near the top of the troposphere, particularly in the western Pacific and Indian Oceans. The phenomenon is named for its founders. The period varies between 30 and 50 days and is accompanied by strong fluctuations in deep convection often clearly visible from satellites. It is a major contributor to intraseasonal variability in the equatorial regions from eastern Africa to the central Pacific and may play a role in triggering stormy regimes even in the eastern Pacific, the Caribbean, and the Atlantic.

Mean Sea-Level Heights actually vary across the Pacific Basin, nearly 2 feet (23 inches) on average between the western Pacific (the Philippines) and the eastern Pacific (Panama coast) between the phases of the ENSO due to the action of the trade easterlies. These winds push water westward along the equator and pile it up in the western areas. In El Niños, as the trade easterlies weaken, water sloshes eastward and sea levels rise in the eastern Pacific and fall in the western Pacific. This drop in sea level can cause exposure damage to the sensitive coral in the western Pacific.

The Climate Diagnostics Center developed the **Multivariate ENSO Index** (MEI) to provide a new comprehensive data set incorporating multiple factors that differ in the two phases of ENSO, including air temperatures, sea surface temperatures, sea-level pressure, surface wind, and cloudiness. The hope was that this reference data set would become the standard for all researchers, providing the most consistent and generally useful results.

North Atlantic Oscillation (NAO) is a north–south oscillation of pressure in the North Atlantic. It is one of the most dominant modes of global climate variability with widespread influence, especially in winter. In winter, pressures are normally low to the north near Iceland (called the Iceland Low) and high to the south around the Azores (called the Azores High) off the coast of Africa. At times, one or both of these pressure systems are stronger than normal. The result is an enhanced westerly flow in between the two features across the Atlantic. This tends to drain cold air off of North America, and drive mild Atlantic air into Europe. The result is milder than normal weather in both the United States and Europe.

At other times, high pressure develops or moves into the far northern Atlantic. When this happens, pressures fall to the south resulting in a flip-flop of the normal pressure and wind flow patterns across the Atlantic. This mode of the NAO results in cold weather in both the United States and Europe. This is because cold air gets bottled up over the United States while in Europe, cold air from Eastern Europe and Asia builds westward sometimes all the way to the Atlantic Coast.

Winters when the strong Iceland Low and Azores High dominate (e.g., 1982–1983, 1988–1989, 1989–1990, and 1994–1995), are typically mild and often snowless winters. Winters when the flip-flop pressure pattern dominates (1916–1917, 1935–1936, 1962–1963, 1968–1969 and 1995–1996) are on the other hand typically cold and snowy. During the winter of 1995–1996, many new all-time records were set for seasonal snowfall in the major cities of the eastern and central United States.

The NAO was in the "cold" mode for the period from the 1940s to the 1960s and then in the "warm mode" from around 1970 to the mid-1990s. A rapid change to the cold mode in 1995–1996 may signal the start of the cold cycle.

The NAO is believed related to a huge circulation in the ocean called the thermohaline circulation, which goes through decadal scale changes. This gives us hope that coupled atmospheric–oceanic models may some day provide seasonal outlooks for the NAO.

Pacific Decadal Oscillation is a recently discovered long-term oscillation in the Pacific oceanic circulation and associated warm and cool water pools that may have a major impact on both global weather patterns and the relative frequency of El Niño and La Niña. Jet propulsion laboratory scientists believe we have entered a cool water phase in the eastern Pacific similar to what was observed several decades ago. Historical data suggests warmer than normal water dominated in the eastern Pacific from 1925 to 1945. A cool phase then developed and lasted from 1947 to 1977 followed by a warm phase from 1977 to 1997.

A cool phase favors more La Niñas and an increased risk of summer heat waves and droughts. The cooling of the water however may boost fish populations in the Pacific Northwest. This is especially true for the salmon populations, which had dropped dramatically after the warming in 1977. In the warm phase, El Niños were more favored.

The **Pacific North America Index (PNA)** is a measure of one of the primary modes of global climate variability. It measures pressures relative to normal in the upper atmosphere at jet stream level across a large area from the eastern Pacific to eastern North America. This steering flow pattern and associated surface weather features vary day to day. However, because of the different thermal characteristics of the land and ocean, the large north to south mountain barrier(s) which interferes with the flow pattern and the earth's rotation on its axis, the upper flow tends to conform to one of two general modes, the two modes of the PNA.

In one mode of the PNA, pressures are below normal and the jet stream dips to the south in the eastern Pacific, then turns north as it passes over the western mountains and then turns south again east of the Rocky Mountains. When it is very amplified in winter, this mode can deliver very cold air to the eastern two thirds of the nation. This mode is favored in El Niño.

In the other mode of the PNA, pressures are above normal off the west coast pushing the jet stream north of its normal position. Pressures then tend to be below normal in the western United States with the polar jet stream depressed south of normal. The jet then turns north east of the Rocky Mountains. This pattern brings the coldest weather to the western mountains and mild conditions to the southeastern states. This mode is favored in La Niña winters.

The **Peru Current (Humboldt Current)** is a relatively slow and shallow cold-water current of the southeast Pacific Ocean. It is the southern hemisphere counterpart to the cold California Current of the North Pacific. It originates as the West Wind Drift of the southern hemisphere that flows toward South America south of 40 degrees south latitude. Some of this cold water flows east through the Drake Passage around the southern end of South America. The rest turns north as a shallow current that parallels the coast as far north as 4 degrees south, before turning westward. The cold temperature of the water in enhanced by upwelling of deep-ocean water caused by drag on the ocean by the southeast trade winds and a turning to the left of the water due to the earth's rotation (Coriolis Force). These factors push water away from the land, thus drawing up cold water from beneath to replace it. The upwelling brings nutrients to the surface making the waters one of the world's greatest fishing grounds.

The Peru current is also called the Humboldt current, named for Alexander von Humboldt, who in 1802 first measured and documented the cold temperature of the water.

The **Quasi-Biennial Oscillation (QBO)** occurs when at very high levels of the atmosphere (30 to 50mb or 10 to 12 miles), the winds tend to oscillate from westerly to easterly and then back to westerly on a cycle that averages about 29 months. In a series of papers in the late 1980s and early 1990s, researchers at the CPC and NCAR found some possible correlations of ENSO weather patterns with phases of the QBO. It is especially true in winter when the expanse of cold polar air (called the polar vortex) and jet streams are farthest south and most sensitive to changes in the equatorial region.

Researchers at NCAR and the CPC also found apparent relationships between winter weather patterns and the QBO and the 11-year solar cycle. Around the peak of the solar cycle and during the quiet years, patterns vary depending on the phase of the QBO. In most areas, the patterns in the westerly phase appear to be nearly the opposite of the favored patterns in the easterly phase. And in most areas, patterns near the solar minimum appear to be nearly the opposite of those at the solar maximum.

First theorized by Carl-Gustav Rossby in the 1930s, **Rossby Waves** are very large planetary waves observed in both the atmosphere and oceans. Although the waves are very easy to observe in the meandering of the jet stream in the atmosphere, they were more difficult to find in the oceans until the advent of satellite oceanography. In the ocean they travel from east to west paralleling the equator. They are several hundreds of kilometers in length, travel at just a few kilometers per day. They raise sea level by 10 to 20 centimeters and lower the thermocline by tens of meters. They may take 9 months or more to cross the Pacific. Although Kelvin waves can initiate an El Nino, the Rossby waves, which are generated when the Kelvin waves encounter the South American Coast, are said to begin the process of an El Nino's decay.

Sea Surface Temperature Anomalies are departures from normal of ocean temperatures. Usually measured at the surface by ships, fixed buoys, satellite and even aircraft. Sea surface temperature anomalies in the tropical Pacific accompany both El Niño and La Niña and are typically opposite in sign. In El Niño, the water in the eastern and central Pacific is warmer than normal while water in the western Pacific tends to be colder than normal. In La Niñas, the eastern and central tropical Pacific waters are colder than normal while waters in the western Pacific are warmer than normal.

The **Southern Oscillation** is a interannual see-saw in sea-level pressure across the tropical Pacific closely linked with El Niño and La Niña. It was first documented and named by Sir Gilbert Walker in the 1930s. He noted there was an inverse relationship between surface air pressures at two sites: Darwin, Australia and the South Pacific Island of Tahiti. High pressure at one site is almost always concurrent with low pressure at the other and vice versa. It represents a standing wave or "see-saw", a mass of air oscillating back and forth in the tropical and subtropical Pacific.

The **Southern Oscillation Index (SOI)** is a measure of the state of the Southern Oscillation first proposed by Walker. It measures the pressure difference between Darwin, Australia and the South Pacific Island of Tahiti. In Walker's SOI, the pressures at both location are "normalized" (relative to normal) in order to account for normal seasonal differences.

In El Niños, unusually high pressure typically develops in the western Tropical Pacific, whereas pressures in the eastern Tropical Pacific become unusually low (a negative SOI). This causes a weakening in the trade winds, which can reduce the cool water upwelling. This leads to a subsequent warming of the waters in the eastern Tropical Pacific. These changes in pressures, winds, and water temperatures cause the tropical showers that normally are found in the western Tropical Pacific to shift east to the east central Pacific and in extreme events all the way to the coast of Peru.

In La Niñas, pressures rise in the eastern Pacific while they fall in the western Pacific (a positive SOI). This causes an increase in the trade winds, enhanced upwelling, and a cooling of the waters in the eastern Pacific. These changes cause a suppression of showers in the eastern Tropical Pacific and a further enhancement of the normal rainfall in the western Tropical Pacific.

TAO/TRITON is an array of approximately 70 moorings, that measure ocean temperatures to a depth of 500 meters over the entire equatorial Pacific. This data, together with data from satellites, are used to monitor in real time, the changes in the ocean that accompany the ENSO cycle.

A **Teleconnection** is a linkage between weather features or changes occurring in widely separated regions of the globe. A teleconnection often takes the form of a statistical relationship (correlation) between a parameter in one location and the parameter in other locations. This is important with regards to ENSO because anomalies in atmospheric pressure associated with the Southern Oscillation in one part of the Tropical Pacific are usually associated with anomalies of the opposite sign in the other side of the Tropical Pacific and with specific anomalies in pressure (and other atmospheric conditions) in other regions of the globe.

The **Thermocline** is the oceanic equivalent of the inversion in the atmosphere. It is the boundary between well-mixed warm water near the surface and deeper colder water. It is normally about 40 meters (130 feet) deep in the eastern Pacific but varies between 100 and 300 meters (330 to 660 feet) deep in the western Pacific. The thermocline varies considerably across the Tropical Pacific in the El Niño and La Niña. In ENSO events, the thermocline is directly proportional to the departure from normal of sea level. A deeper than normal thermocline is usually associated with above normal sea levels. A shallower than normal thermocline meanwhile is typically associated with below normal sea levels

In El Niños, the thermocline is much deeper than normal in the eastern Pacific (the warm water layer is deeper than normal) while it is usually shallower than normal in the western Pacific. In La Niñas, the thermocline is shallower than normal in the Eastern Pacific and deeper than normal in the west.

The depth of the thermocline is affected by Kelvin waves (which move east and increase the thermocline depth), and by Rossby waves (which move west and usually cause the thermocline depth to diminish).

The **Tropical Ocean Global Atmosphere (TOGA)** program was a 10-year research project that resulted in major strides towards an understanding of the ENSO phenomenon. It demonstrated the feasibility operational multi-season climate prediction of ocean temperatures based on numerical models and clarified the nature of the remote, planetary scale, atmospheric response to these anomalies.

Trade Winds are generally the most dependable and steadiest winds on the earth. The trades are found in the subtropics between a belt of clear skies and high pressure called the subtropical high (centered in the oceans around 30 to 40 N and S) and the zone of low pressure, clouds, and showers near the equator called the Intertropical Convergence Zone. The trade winds blow from the northeast in the northern hemisphere and southeast in the southern hemisphere. The trade winds are often referred to as the easterlies.

In La Niñas, the trade winds blow stronger than normal, which enhances the cold water upwelling near the South American coast and carries the resulting cold surface water westward along the equator. These strong trades also act to pile up warm water in the western Pacific.

In El Niños, the trade winds weaken and may reverse in the western Pacific. This allows warm water to "slosh" eastward in the Pacific Basin. The weakened trade winds reduce the upwelling of cold water near the South American coast and water warms at the surface. When the warm water arrives from the west, sea levels rise and the depth of the warm water layer near the surface deepens. The warm water appears as a characteristic plume extending west along the equator to near or past the International Dateline.

The trade winds from both hemispheres converge on the equatorial region resulting in **Tropical Easterlies**. The easterlies tend to be strongest in cold La Niña events. The easterlies can reverse and become westerlies in warm El Niño events, at least in the western half of the Pacific Basin. The strength of the easterlies is monitored closely as changes might signal a change of phase of ENSO.

Upwelling is the rising of cold water from the deeper ocean to the surface caused by the removal of surface water by the wind. When the action of the wind causes surface water near the land to move away from the coast, it must be replaced by water from beneath. Because deeper water is usually colder water, the result is that the upwelling produces a cooling of the surface. The trade winds near the South American coast act to produce upwelling. This upwelling is enhanced in La Niñas because the trade winds are stronger than normal. The upwelling is diminished in El Niños as trade winds diminish.

The **Walker Circulation** is another three-dimensional circulation in the Tropical Pacific associated with El Niño and La Niña. It was proposed by the renowned meteorologist Jacob Bjerknes in 1969 and named in honor of Walker. In El Niño, the air sinks in the western areas, moves eastward, and rises in the central (and in stronger El Niños the eastern) Pacific only to return westward at higher altitudes. In La Niña, the pattern reverses as air sinks in the eastern Pacific and moves westward to rise in the western Pacific.

A **Warm Event** (or **Warm Phase of ENSO** or **ENSO Warm Event**) is an El Niño or an anomalous warming of the waters in the eastern and central Tropical Pacific. Warming of the waters there is often accompanied by a relative cooling of the waters in the western tropical Pacific.

Westerly Wind Bursts are short duration (one to several days) low-level wind events along and near the equator in the western Pacific and Indian Ocean. It is most common in El Niño years from September to January and normal (neutral ENSO) years from October to December. It is often absent in La Niña years.

Bibliography

Allan, R. J., Lindesay, J., and Parker, D. (1996). El Niño Southern Oscillation and climatic variability. *El Niño Southern Oscillation and Climatic Variability*. Perth, Australia: Commonwealth Scientific and Research Organization, 416 pp.

Allan, R. J.; Nicholls, N., Jones, P. D., and Butterworth, I. J. (1991). A further extension of the Tahiti–Darwin SOI, early ENSO events and Darwin pressure. *Journal of Climate* 4, no. 7, pp. 743–749.

Barnett, T. P. (1984). Prediction of the El Niño of 1982–3. *Monthly Weather Review* 112, no. 7, pp. 1403–1407.

Barnett, T. P. (September 11, 1997). Testimony to Congress on how climate forecasts are being used. *Congressional Record*.

Barnston, A. G., and Livezey, R. E. (1987). Classification, seasonality and persistence of low-frequency atmospheric circulation patterns. *Monthly Weather Review* 115, pp. 1083–1126.

Bigg, G. R. El Niño and the Southern Oscillation. (January 1990). *Weather* 45, pp. 2–8.

An easily understood look at the workings of El Niño, 1990.

Bjerknes, J. (1966). A possible response of the atmospheric Hadley circulation to equatorial anomalies of ocean temperatures. *Tellus* 18, pp. 820–829.

Bjerknes, J. (1969). Atmospheric teleconnections from the equatorial Pacific. *Monthly Weather Review*, 97, pp. 163–172.

Bjerknes, J. (1972). Large-scale atmospheric response to the 1964–65 Pacific equatorial warming. *Journal of Physical Oceanography* 2, pp. 212–217.

Bove, M. C., Elsner, J. B., Landsea, C. W., Niu, X., O'Brien, J. J. (1998).

Effects of El Niño on U.S. landfalling hurricanes. *Revisited* 70, no. 11, pp. 2477–2482.

Bradley, R. S., Diaz, H. F., Kiladis, G. N., and Eischeid, J. K. (1987), ENSO signal in continental temperature and precipitation records. *Nature* 327, no. 6122, pp. 497–501.

Brock, R. G. (1984). El Niño and world climate: piecing together the puzzle. *Environment* 26, no. 3, pp. 14–20, 37–39.

Brooks, C. E., and Brady, H. W. (1921). The clash of the Trades in the Pacific. *Quarterly Journal of the Royal Meteorological Society* 47, pp. 1–13.

Canby, T. Y. (February 1984). El Niño's ill wind. *National Geographic*, pp. 144–183.
An excellent overview on the phenomenon, explaining El Niño within the time frame of the strongest event in modern times. Chronicles the effects worldwide through use of photographs and schematics.

Cane, M. A., and Zebiak, S. E. (1985). A theory for El Niño and the Southern Oscillation. *Science* 228, no. 4703, pp. 1085–1087.

Caviedes, C. N. (1991). Five hundred years of hurricanes in the Caribbean: their relationship with global climatic variabilities.
It is found that lesser numbers of hurricanes occur during El Niño conditions in the tropical Pacific and increased numbers tend to develop during Anti-Nino eposides. *GeoJournal* 23, no. 4, pp. 301–10.

Changnon, S. A. (1999). Impacts of the 1997–98 El Niño—generated weather in the United States. *Bulletin of the American Meteorological Society* 80, no. 9, pp. 1819–1827.

Deser, C., and Wallace, J. M. (1987). El Niño events and their relation to the Southern Oscillation: 1925–86. *Journal of Geophysical Research* 92, no. C13, pp. 14189–14196.

Deser, C., and Wallace, J. M. (1990). Large-scale atmospheric circulation features of warm and cold episodes in the tropical Pacific. *Journal of Climate* 3, no. 11, pp. 1254–1281.

Enfield, D. B. (1987); Progress in understanding El Niño. *Endeavour, New Series* 11, no. 4, pp. 197–204.

Fairbridge, R. W. (1990). Solar and lunar cycles embedded in the El Niño periodicities. *Cycles* 41, no. 2, p. 65.

Fraedrich, K., Muller, K., and Kuglin, R. (1992). Northern hemisphere circulation regimes during the extremes of the El Niño/Southern Oscillation. *Tellus, Series A (Dynamic Meteorology and Oceanography)* 44A, no. 1, pp. 33–40.

Glantz, M. H. (1984). Floods, fires, and famine: is El Niño to blame? *Oceanus* 27, no. 2, pp. 14–19.

Glantz, M. H. (1996). *Currents of Change: El Niño's Impact on Climate and Society.* Cambridge: Cambridge University Press.

Graham, N. E., and White, W. B. (1988). The El Niño cycle: a natural oscillator of the Pacific Ocean—atmosphere system. *Science* 240, no. 4857, pp. 1293–1302.

Gray, W. M. (1984). Atlantic seasonal hurricane frequency. I. El Niño and 30 mb quasi-biennial Oscillation influences. *Monthly Weather Review* 112, no. 9, pp. 1649–1668.

Handler, P., and Handler, E. (1983). Climatic anomalies in the tropical Pacific Ocean and corn yields in the United States. *Science* 220, no. 4602, pp. 1155–1156.

Association between El Niño events and annual corn production in the United States was investigated.

Hirono, M. (1988). On the trigger of El Niño Southern Oscillation by the forcing of early El Chichon volcanic aerosols. *Journal of Geophysical Research* 93 D5, pp. 5365–5384.

Kiladis, G. N., and Diaz, H. F. (1989). Global climate anomalies associated with extremes of the Southern Oscillation. *Journal of Climate* 2, pp. 1069–1090.

Kirchner, I., and Graf, H. F. (1995). Volcanos and El Niño: signal separation in Northern Hemisphere winter. *Climate Dynamics* 11, no. 6, pp. 341–358.

Kocin, P. J., and Uccellini, L. W. (1990). Snowstorms along the northeastern coast of the United States: 1955 to 1985. *Meteorological Monograph* 44, 280 pp.

Kousky, V. E., Kagano, M. T., and Cavalcanti, F. A. (1984). A review of the Southern Oscillation: oceanic-atmospheric circulation changes and related rainfall anomalies. *Tellus, Series A (Dynamic Meteorology and Oceanography)* 36A, no. 5, pp. 490–504.

Kuo, H. L. (1989). Long-term oscillations in the coupled atmosphere-ocean system and El Niño phenomenon. *Journal of Climate* 2, no. 12, p. 1421.

Leetmaa, A. (1989). The interplay of El Niño and La Niña. *Oceanus* 32, no. 2, p. 30.

Legler, David M., Kelly, J. Bryant, and O'Brien, James J. (1997). Impact of ENSO related climate anomalies on crop yields in the US. submitted to climatic change July 18, 1997. *Abstract Online http://www.coaps.fsu.edu/legler/Leg97–1sum.html*

Mantua, N. J., Hare, S. R., Zhang, Y., Wallace, J. M., and Francis, R. C. (1997). A Pacific interdecadal climate oscillation with impacts on salmon production. *Bulletin of the American Meteorological Society* 78, pp. 1069–1079.

McCreary, J. P., Jr., and Anderson, D.L.T. (1984). A simple model of El Niño and the Southern Oscillation. *Monthly Weather Review* 112, no. 5, p. 934–946.

Mo, K. C., and Kalnay, E. (1991). Impact of sea surface temperature anomalies on skill of monthly forecasts. *Monthly Weather Review* 119, no. 12, pp. 2771–2793.

Mock, D. R. (1981). *The Southern Oscillation: Historical Origins.* Seattle: University of Washington.

Namias, J., and Cayan, D. R. (1984), El Niño: implications for forecasting (USA). *Oceanus* 27, no. 2, pp. 41–47.

Nicholls, N. (1988). El Niño—Southern Oscillation impact prediction. *Bulletin—American Meteorological Society* 69, no. 2, pp. 173–176.

Nicholls, N. (1990). Low-latitude volcanic eruptions and the El Niño/ Southern Oscillation: a reply. *International Journal of Climatology* 10, no. 4, pp. 425–429.

O'Brien, J. J., Richards, T. S., and Davis, A. C. (April 1996). The effect of El-Niño on U.S. landfalling hurricanes, *Bulletin of the American Meteorological Society* 77, no. 4, pp. 773–774. Available Online <http://www.coaps.fsu.edu/richards/paper.html

Oceanus. 27 no. 2. The entire summer 1984 issue was devoted to El Niño Overall, an excellent summary of the phenomenon and its ramifications. This is one issue worth hunting down. This issue included the following papers:
  • Arntz, W. E. El Niño and Peru: positive aspects.
  • Glantz, M. H. Floods, fires, and famine: is El Niño to blame?
  • Harrison, D. E., and Cane, M. A. Changes in the Pacific during the 1982–83 event (El Niño).
  • McGowan, J. A. The California El Niño, 1983.
  • Namias, J., and Cayan, D. R. El Niño: implications for forecasting (USA).
  • Rasmusson, E. M. El Niño: the ocean/atmosphere connection.
  • Toole, J. M. Sea Surface Temperature in the Equatorial Pacific.
  • Webster, F. Studying El Niño on a global scale (TOGA).

Orlove, B. S., Chiang, J. C., and Cane, M. A. (January 6, 2000). Forecasting Andean rainfall and crop yield from the influence of El Niño on Pleiades visibility. *Nature* 43, pp. 68–71.

Piechota, T. C., and Dracup, J. A. (May 1996). Drought and regional hydrologic variation in the United States: associations with the El Niño-Southern Oscillation. *Water Resources Research* 32, no. 5, pp. 1359–1370.
  Drought and stream flow in the Pacific northwest and southeastern United States are strongly associated with extreme phases of ENSO.

Pielke, Roger A., and Landsea, Christopher N. (1999). La Niña, El Niño, and Atlantic hurricane damage in the United States. *Bulletin of the American Meteorological Society* 80, no. 10, pp. 2027–2033.

Portman, D. A., and Gutzler, D. S. (1996). Explosive volcanic eruptions, the El Niño-Southern Oscillation, and U.S. climate variability. *Journal of Climate* 9, no. 1, pp. 17–33.

Rasmusson, E. M. (1984). El Niño: the ocean/atmosphere connection. *Oceanus* 27, no. 2, pp. 5–12.

Rasmusson, E. M. (1985). El Niño and variations in climate. *American Scientist* 73, no. 2, pp. 168–77.

Rasmusson, E. M., and Wallace, J. M. (1983). Meteorological aspects of the El Niño/Southern Oscillation. *Science* 222, no. 4629, pp. 1195–2020.

Reynolds, R. W., and Smith, T. M. (1994). Improved global sea surface temperature analyses using optimum interpolation. *Journal of Climate.* 7, pp. 929–948.

Robock, A., Taylor, K. E., Stenchikov, G. L., and Yuhe Liu. (1995). GCM evaluation of a mechanism for El Niño triggering by the El Chichon ash cloud. *Geophysical Research Letters* 22, no. 17, pp. 2369–2372.

Ropelewski, C. F., and Halpert, M. S. (1986). North American precipitation and temperature patterns associated with the El Niño/Southern Oscillation (ENSO). *Monthly Weather Review* 114, no. 12, pp. 2352–2362.

Ropelewski, C. F., and Halpert, M. S. (1987). Global and regional scale precipitation patterns associated with the El Niño/Southern Oscillation. *Monthly Weather Review* 115, no. 8, pp. 1606–1626.

Ropelewski, C. F., and Jones, P. D. (1987) (An extension of the Tahiti–Darwin Southern Oscillation Index. *Monthly Weather Review* 115, pp. 2181–2165.

Rothstein, L. M. (Fall–Winter 1996). The El Niño Southern Oscillation phenomenon seeking its trigger and working toward prediction. *Oceanus* 39, no. 2, pp. 39–41.

Schell, I. I. (1956). On the nature of the Southern Oscillation. *Journal of Meteorology* 13, pp. 595–596.

Schell, I. I. (1965). The origin and possible prediction of the fluctuations in the Peru Current and upwelling. *Journal of Geophysical Research* 70, pp. 5529–5540.

Stevens, W. K. (January 3, 1989). Scientists link '88 drought to natural cycle in Tropical Pacific. *New York Times*, 7(B).

Trenberth, K. E. (1991). *General Characteristics of the El Niño Southern Oscillation, Teleconnections Linking Worldwide Climate Anomalies.* Cambridge: Cambridge University Press.

Troup, A. J. (1965). The Southern Oscillation. *Quarterly Journal of the Royal Meteorological Society* 102, pp. 490–506.

University of Hawaii Sea Level Center (UHSLC). (2000). *Biographical Summary Klaus Wyrtki.* Available on-line from http://www.soest. hawaii.edu/WCCRP/KW.html

Vallis, G. K. (1988). Conceptual models of El Niño and the Southern Oscillation. *Journal of Geophysical Research* 93, no. 11, p. 13, 979.

Von Storch, H., Sundermann, J., and Magaard, L. (February 25, 1999).

Interview with Klaus Wyrtki, GKSS 99/E/74, 44 pp. Available from GKSS-Forschungszentrum. Geesthacht GmbH, Bibliothek, Postfach 11 60, D-21494 Geesthacht, Germany, Fax: (49) 04152/871717.

Walker, D. A. (January 24, 1995). More evidence indicates link between El Niño and seismicity. *EOS* 76, pp. 1, 34, 35.

Walker, G. T. (1928). World weather. *Monthly Weather Review.* 56, pp. 167–170.

Wallace, J. M., and Vogel, S. (Spring 1994). *El Niño and Climate Prediction, Reports to the Nation on Our Changing Planet.* UCAR and NOAA Office of Global Programs.
This report is available from UCAR Office of Interdisciplinary Earth Studies, P.O. Box 3000, Boulder, CO 80307–3000 (303) 497–2692.

Webster, F. (1984). Studying El Niño on a global scale (TOGA). *Oceanus* 27, no. 2, pp. 58–62.

White, W. B., Pazan, S. E., and Inoue, M. (1987). Hindcast/forecast of ENSO events based upon the redistribution of observed and model heat content in the western tropical Pacific, 1964–86. *Journal of Physical Oceanography* 17, no. 2, pp. 264–280.

Willet, H. C, and Bodurtha, F. T. (1952). An abbreviated Southern Oscillation. *Bulletin of the American Meteorological Society.* 33, pp. 429–430.

Wolter, K., and Timlin, M. (September 1998). Measuring the strength of ENSO events, how does 1997/98 Rank? *Weather* 3, no. 9, pp. 315–324.

Wyrtki, K. (1975). El Niño—The dynamic response of the equatorial Pacific Ocean to atmospheric forcing. *Journal of Physical Oceanography* 5, no. 4, pp. 572–584.

Wyrtki, K. (1982). The Southern Oscillation, ocean–atmosphere interaction and El Niño. *Marine Technology Society Journal* 16, no. 1, pp. 3–10.

Wyrtki, K., Stroup, E., Patzert, W., Williams, R., and Quinn, W. (1976). Predicting and observing El Niño. *Science* 191, no. 4225, pp. 343–346.

Zebiak, S. E. (1989). Oceanic heat content variability and El Niño cycles. *Journal of Physical Oceanography* 19, no. 4, p. 475.

FURTHER READING: THE BASIC LIST

This list contains the authors' recommendations for publications that would provide the reader a good basic overall understanding of ENSO and its many ramifications.

Aceituno, P. (April 1993). El Niño, the Southern Oscillation and ENSO: confusing names for a complex ocean–atmosphere interaction. *Bul-*

letin of the American Meteorological Society 73, pp. 483–485.
A brief review of the history and nomenclature of El Niño and the Southern Oscillation.

Arnold, C. (1998). *El Niño: Stormy Weather for People and Wildlife.* New York: Clarion Books.

Barber, R. T., and Chavez, F. P. (1983). Biological consequences of El Niño. *Science* 222, no. 4629, pp. 1203–1210.
An account of the many environmental effects that El Niño, especially strong El Niños, can bring. Focuses on the Super El Niño of 1982–1983. Provides oceanic cross-sections of nitrates, chlorophyll, etc., during the event.

Barnett, T., Graham, N., Cane, M., Zebiak, S., Dolan, S., O'Brien, J., and Legler, D. (1988). On the prediction of the El Niño of 1986–1987. *Science* 241, no. 4862, p: 192–196. An historical evaluation done during the TOGA research decade, including a study of the effect on worldwide weather patterns.

Blueford, J. (August 1988). El Niño: ocean temperature affects weather and life cycles. *Instructor* 98, pp. 74–77.
A science teacher's approach to explaining El Niño. A good listing of historical events.

Changnon, S. A. (1999). Impacts of the 1997–98 El Niño—generated weather in the United States. *Bulletin of the American Meteorological Society* 80, no. 9, pp. 1819–1827.
An interesting assessment of both the benefits and costs of the 1997–98 event. The media tend to focus on the negative, but as the author shows, sometimes the ignored benefits more than outweigh the costs.

Diaz, H. F., and Markgraf, V. (1993). *El Niño: historical and paleoclimatic aspects of the Southern Oscillation.* 476 pp. Cambridge: Cambridge University Press.
Two authors well known in climate circles look back at the phenomenon over history. The authors use proxy data (such as tree ring, marine sediment records, and South American fishery records) to evaluate the ENSO state before modern records.

Diaz, H. F., and Markgraf, V., eds. (2000). *El Niño and the Southern Oscillation: Multi-scale Variability and Global and Regional Impacts.* 49 pp. Cambridge: Cambridge University Press. A new comprehensive review of ENSO and its many ramifications.

Gantenbein, D. (1995). El Niño the weathermaker. *Popular Science* 246, pp. 76–78.

Glantz, M. H. (1996). *Currents of Change: El Niño's Impact on Climate and Society.* Cambridge: Cambridge University Press.
A prolific and well-respected author on the phenomenon catalogues the effects on world climates and on society.

Green, P. M. Legler, D. M. Miranda V. C. J. and O'Brien, J. J. (1997).

The North American Climate Patterns Associated with the El Niño–Southern Oscillation. Available Online http://www.coaps.fsu.edu/lib/booklet/
The authors present graphics of the temperature and precipitation anomalies typically associated with each of the four seasons during mature El Niño and La Niña events.

Nicholls, N. (1988). El Niño—Southern Oscillation impact prediction. *Bulletin of the American Meteorological Society* 69, no. 2, pp. 173–176. An interesting study on the many impacts of El Niño, including the effects on marine life.

Philander, S.G.H. (1989). *El Niño, La Niña and the Southern Oscillation.* San Diego: Academic Press, 294 pp.
A definitive college level text. A comprehensive summary of the two phases of the Southern Oscillation by one of the most widely read and popular authors on the phenomena. A paper by the author, El Niño and La Niña, can be found in *American Scientist Magazine 77*, no. 5 (1989), pp. 451–459.

Philander, S.G.H. (1983). El Niño Southern Oscillation phenomena (Ecuador, Peru). *Nature* 302, no. 5906, pp. 295–301.

Philander, S.G.H. (1989). El Niño and La Niña. A vast system of ocean–atmosphere exchanges covering the Tropical Pacific *American Scientist* 77, no. 5, p. 451.

Ramage, C. S. (1986). El Niño. *Scientific American*, 254, pp. 76–83. A look at the far-reaching climate disturbances caused by the phenomenon.

Rasmusson, E. M., and Hall, J. M. (August 1983). El Niño: The great Equatorial Pacific Ocean warming event of 1982–1983. *Weatherwise* 36, pp. 166–175.
A meteorological review of the Super El Niño of 1982–1983.

Robinson, G. R. (1987). Negative effects of the 1982–83 El Niño on Galapagos marine life. *Oceanus*, 30, no. 2, pp. 42–48.
A funny, yet sad, article that is worth searching for. Focuses on the effects on life close to the heart of the El Niño warm plume.

Ropelewski, C. F., and Halpert, M. S. (1986). North American precipitation and temperature patterns associated with the El Niño/Southern Oscillation (ENSO). *Monthly Weather Review* 114, no. 12, pp. 2352–2362.
A review of the ENSO effects on temperature and precipitation, focusing on North America.

Ropelewski, C. F., and Halpert, M. S. (1987). Global and regional scale precipitation patterns associated with the El Niño/Southern Oscillation. *Monthly Weather Review* 115, no. 8, pp. 1606–1626.
Ground-breaking study that demonstrated how truly global the climate changes were resulting from ENSO. Authors showed the areas globally that were unusually wet or dry during the phases of the El Niño/Southern Oscillation.

Thayer, V. G., and Barber, R. T. (October 1984). At sea with El Niño: the fish were gone, the birds were gone and the winds blew in the wrong direction. *Natural History* 93, pp. 4–12.

A good summary of the effects on nature brought about by the ocean and atmospheric extremes resulting from El Niño. El Niño experienced from the inside. Features a nice narrative following the researchers in their work.

Trenberth, K. E. (1990). Recent observed interdecadal climate changes in the Northern Hemisphere. *Bulletin of the American Meteorological Society* 71, no. 7, pp. 988–993.

North Pacific changes appear to be linked through teleconnections to tropical atmosphere–ocean interactions and the frequency of El Niño.

Wallace, J. M., and Vogel, S. (Spring 1994). *El Niño and Climate Prediction, Reports to the Nation on Our Changing Planet.* UCAR and NOAA Office of Global Programs.

This report is available from UCAR Office of Interdisciplinary Earth Studies, P.O. Box 3000, Boulder, CO 80307–3000 (303) 497–2692. An excellent summary of the history, causes, and many impacts of the great El Niño of 1982–1984.

Weare, B. C. (1996). An extension of an El Niño index. *Monthly Weather Review* 114, no. 3, pp. 644–647.

Willet, H. C., and Bodurtha, F. T. (1952). An abbreviated Southern Oscillation. *Bulletin of the American Meteorological Society* 33, pp. 429–430.

Wyrtki, K. (1975). El Niño—the dynamic response of the equatorial Pacific Ocean to atmospheric forcing. *Journal of Physical Oceanography* 5, no. 4, pp. 572–584.

A classic study by one of the pioneers in ENSO research outlining how the atmosphere and oceans work together as a coupled system during the oscillation we know as the Southern Oscillation.

Yang, X. B., and Scherm, Harold. (1997). El Niño and infectious disease. *Science* 275, p. 739.

One-page article discussing the impacts of El Niño on diseases.

Zebiak, S. E. (May 31, 1985). A theory for El Niño and the Southern Oscillation. *Science* 228, pp. 1085–1087.

A brief (two-page) paper on modeling El Niño (here with only the essential physics) in an attempt to better understand its reoccurrences.

FURTHER READING—A LIST BY CATEGORY

The following more complete listing of ENSO papers, books, and articles was compiled with assistance from COAPS http://www.coaps.fsu.edu/

lib/ and NOAA's Climate Diagnostics Center http://www.cdc.noaa.gov/
~cas/ENSO.pubs.html.

ENSO and Drought

Andrade, E. R., Jr., and Sellers, W. D. (1988). El Niño and its effect on
    precipitation in Arizona and western New Mexico. *Journal of Cli-
    matology* 8, no. 4, pp. 403–410.

Armstrong, L. and Veomett, E. (September, 29, 1997) An ill wind for busi-
    ness: El Niño's mischief could set off a chain of economic calamities.
    Business Week, no. 3546, pp. 32–33.

Arpe, K. (2000). Connection between Caspian Sea level variability
    and ENSO. *Geophysical Research Letters* 27, no. 17, pp. 2693–
    2696.

Bell, A. (1986). El Niño, and prospects for drought prediction. *CSIRO
    Environmental Research* 49, pp. 12–18.

Bhalme, H. N., and Jadhav, S. K. (1984). The Southern Oscillation and its
    relation to the monsoon rainfall. *Journal of Climatology* 4, no. 5,
    pp. 509–520.

Bhalme, H. N., Sikder, A. B., and Jadhav, S. K. (1990). Coupling between
    the El Niño and planetary-scale waves and their linkage with the
    Indian monsoon rainfall. *Meteorology and Atmospheric Physics* 44, no.
    1–4, pp. 293–305.

Bhalme, H. N., Sikder, A. B., and Jadhav, S. K. (1990). Relationships be-
    tween planetary-scale waves, Indian monsoon rainfall and ENSO.
    Mausam 41, no. 2, pp. 279–84.

Bonsal, B. R., and Lawford, R. G. (2000). Teleconnections between El
    Niño and La Niña events and summer extended dry spells on the
    Canadian prairies. *International Journal of Climatology* 19, no. 13,
    pp. 1445–1458.

Brahmananda Rao, V., Satyamurty, P., Ivaldo, J., and De Brito, B. (1986).
    On the 1983 drought in north-east Brazil. *Journal of Climatology* 6,
    no. 1, pp. 43–51.
    Rainfall anomalies over northeast Brazil are related to the Southern
    Oscillation and SST anomalies in the Tropical Atlantic Ocean.

Brenner, James. (1999), *Southern Oscillation Anomalies and Their Relation
    to Florida Wildfires*. Florida Department of Agriculture Consumer
    Services.

Bunkers, M. J., Miller, J. R., Jr., and DeGaetano, A. T. (1996). An exami-
    nation of El Niño–La Niña–related precipitation and temperature
    anomalies across the Northern Plains. *Journal of Climate* 9, no. 1,
    pp. 147–60.

Burnham, L. (March 1989). The summer of 1988: a closer look at last year's
    drought. *Scientific American*, p. 21.

Dayton, L. (December 7, 1991). Forests and rivers dying in Australian drought. *New Scientist* 132, p. 14.

Dudley, N. J., and Hearn, A. B. (1993). El Niño effects hurt Namoi irrigated cotton growers, but they can do little to ease the pain. *Agricultural Systems* 42, no. 1–2, pp. 103–126.

El Niño and Hawaiian drought. (May 12, 1992). *Weekly Weather and Crop Bulletin* 79, p. 10.

Fleer, H., Schweitzer, B., and Raatz, W. (1984). Rainfall fluctuations in India and Sri Lanka and large-scale rainfall anomalies. *Mausam*, 35, no. 2, pp. 135–144.

There appears to be a linkage between the recurrent Indian droughts and the El Niño phenomena along the west coast of South America.

Francou, B. (1985). "El Niño" and the drought in the High Central Andes of Peru and Bolivia. *Bulletin—Institut Francais des Etudes Andines* 14, no. 1–2, pp. 1–18.

Gordon, H. B., and Hunt, B. G. (1991). Droughts, floods, and sea-surface temperature anomalies: a modelling approach. *International Journal of Climatology*, 11, no. 4, pp. 347–365.

Three different sea-surface temperature anomaly patterns were explored, representative of both El Niño and anti-El Niño events.

Grimm, A. M., Barros, V. R., and Doyle, M. E. (2000). Climate variability in southern South America associated with El Niño and La Niña events. *Journal of Climate* 13, no. 1, pp. 35–58.

Janicot, S., Moron, V., and Fontaine, B. (1996). Sahel droughts and ENSO dynamics. *Geophysical Research Letters* 23, no. 5, pp. 515–518.

Kane, R. P. (1997). Prediction of droughts in north-east Brazil: Role of ENSO and use of periodicities. *International Journal of Climatology* 17, no. 6, pp. 655–665.

Periodicities in rainfall are better short-term predictors of droughts than forecasts based solely on the appearance of El Niño.

Kane, R. P. (2000). El Niño/La Niña relationship with rainfall at Huancayo, in the Peruvian Andes. *International Journal of Climatology* 20, no. 1, pp. 63–72.

Matarira, C. H. (1990). Drought over Zimbabwe in a regional and global context. *International Journal of Climatology* 10, no. 6, pp. 609–625.

Results show significant correlation between the Southern Oscillation and seasonal rainfall over the region.

Maybank, J., Bonsal, B., Jones, K., et al. (June 1, 1995). Drought as a natural disaster. *Atmospheric Ocean* 33, 195–222.

Miao Qilong and Liu Yafang. (1992). Drought, earthquake and El Niño in China. *Earthquake* (Beijing) 2, no. 2, pp. 58–65.

Monastersky, R. (July 23, 1994). Exploiting El Niño to avert African famines. *Science News* 146, p. 52.

Mooley, D. A., and Parthasarathy, B. (1984). Indian summer monsoon and El Niño. *Pure and Applied Geophysics* 121, no. 2, pp. 339–352.

Morliere, A., and Rebert, J. P. (1986). Rainfall shortage and El Niño-Southern Oscillation in New Caledonia, southwestern Pacific. *Monthly Weather Review* 114, no. 6, pp. 1131–1137.

Nicholls, N. (1985). Towards the prediction of major Australian droughts. *Australian Meteorological Magazine* 33, no. 4, pp. 161–166.
The drought index is shown to be closely related to indices of the ENSO and appears to be predictable from changes in tropical Australian air temperature during the preceding summer.

Pearce, F. (March 21, 1992). Drought hits Brazil as climate chaos spreads. *New Scientist* 133, p. 10.

Pereyra Diaz, D., Angulo Cordova, Q., and Palma Grayeb, B. E. (1994). Effect of ENSO on the mid-summer drought in Veracruz State, Mexico. *Atmosfera* 7, no. 4, pp. 211–219.

Phillips, Jennifer, Rajagopalan, Balaji, Cane, Mark, and Rosenzweig, Cynthia. (1999). The role of ENSO in determining climate and maize yield Variability in the U.S. Cornbelt. *International Journal of Climatology*.

Piechota, T. C., and Dracup, J. A. (May 1996). Drought and regional hydrologic variation in the United States: associations with the El Niño-Southern Oscillation. *Water Resources Research* 32, no. 5, pp. 1359–1373.
Drought and stream flow in the Pacific northwest and southeastern United States are strongly associated with extreme phases of ENSO.

Piexoto, J., and Oort A. (1992). *Physics of Climate*. American Institute of Physics, ISBN 0–88318–711–6), 320 pp.

Quinn, W. H., Zopf, David O., Short, Kent S., Kuo Yang, and Richard, T. W. (1981). Historical trends and statistics of the Southern Oscillation, El Niño, and Indonesian droughts (Peru). *Fishery Bulletin* 76, no. 3, pp. 663–678.

Ralph, W. (1984). Drought and the El Niño phenomenon (Australia, Pacific). *CSIRO Environmental Research* 38, pp. 13–15.

Ropelewski, C. F., and Halpert, M. S. (1986). North American precipitation and temperature patterns associated with the El Niño/Southern Oscillation (ENSO). *Monthly Weather Review* 114, no. 12, pp. 2352–2362.

Ropelewski, C. F., and Halpert, M. S. (1987). Global and regional scale precipitation patterns associated with the El Niño/Southern Oscillation. *Monthly Weather Review* 115, no. 8, pp. 1606–1626.

Saseendran, S. A., Rathore, L. S.; and Datta, R. K. (1996). Distribution of monsoon rainfall in India during El Niño associated drought situations. *Annals of Arid Zone* 35, no. 1, pp. 9–16.

Stevens, W. K. (January 3, 1989). Scientists link '88 drought to natural cycle in Tropical Pacific. *New York Times*, pp. 5(B), 7(B).

Thompson, L. M. (1990). Relationship of the El Niño cycle to droughts in the U.S. Corn Belt. *Cycles* 41, no. 1, p. 14.

Trenberth, K., Branstator, G. W., and Arkin, P. A. (December 23, 1988). Origins of the 1988 North American drought. *Science* 242, pp. 1640–1645.

Van Oldenborgh, G. J., Burgers, G., and Tank, A. K. (2000). On the El Niño teleconnection to spring precipitation in Europe. *International Journal of Climatology* 20, no. 5, pp. 565–574.

Wallace, J. M., and Vogel, S. (Spring 1994). *El Niño and Climate Prediction, Reports to the Nation on Our Changing Planet.* UCAR and NOAA Office of Global Programs.
This report is available from UCAR Office of Interdisciplinary Earth Studies, P.O. Box 3000, Boulder, CO 80307–3000, (303) 497–2692.

Weber, G.-R. (1990). North Pacific circulation anomalies, El Niño and anomalous warmth over the North American continent in 1986–1988. *International Journal of Climatology* 10, no. 3, pp. 279–289. Possible causes of the 1988 North American drought.

Webster, F. (1984). Studying El Niño on a global scale (TOGA). *Oceanus* 27, no. 2, pp. 58–62.

Zhu Bingyuan, and Li Donghang. (1992). The relationship between the El Niño events and the drought or excessive rain of Northwest China during 1845 to 1988. *Scientia Atmospherica Sinica* 16, no. 2, pp. 185–192.

ENSO Storm and Flood

Andrade, E. R., Jr., and Sellers, W. D. (1988). El Niño and its effect on precipitation in Arizona and western New Mexico. *Journal of Climatology* 8, no. 4, pp. 403–410.

Bhalme, H. N., and Jadhav, S. K. (1984). The Southern Oscillation and its relation to the monsoon rainfall. *Journal of Climatology* 4, no. 5, pp. 509–520.

Bhalme, H. N., Sikder, A. B., and Jadhav, S. K. (1990). Coupling between the El Niño and planetary-scale waves and their linkage with the Indian monsoon rainfall. *Meteorology and Atmospheric Physics* 44, no. 1–4, pp. 293–305.

Bhalme, H. N., Sikder, A. B., and Jadhav, S. K. (1990). Relationships between planetary-scale waves, Indian monsoon rainfall and ENSO. *Mausam* 41, no. 2, pp. 279–284.

Bove, M. C., Elsner, J. B., Landsea, C. W., Niu, X., and O'Brien, J. J. (1998). Effects of El Niño on U.S. landfalling hurricanes. *Revisited* 70, no. 11, pp. 2477–2482.

Bove, M. C., and O'Brien, J. J. (1997). Impacts of ENSO on United States Tornadic Activity. Available Online <http://www.coaps.fsu.edu/bove/tornado/main.html>

Brownlee, S., and Tangley, L. (October 6, 1997). The wrath of El Niño. *U.S. News & World Report* 123, pp. 16–22.

Caviedes, C. N. (1985). Emergency and institutional crisis in Peru during El Niño 1982–1983. *Disasters* 9, no. 1, pp. 70–74.

Caviedes, C. N. (1991). Five hundred years of hurricanes in the Caribbean: their relationship with global climatic variabilities. *GeoJournal* 23, no. 4, pp. 301–310.
It is found that fewer hurricanes occur during El Niño conditions in the tropical Pacific and increased numbers tend to develop during Anti-Nino episodes.

Chowdhury, A., and Mhasawade, S. V. (1991). Variations in meteorological floods during summer monsoon over India. *Mausam* 42, no. 2, pp. 167–170.
Relationships with the El Niño phenomena are examined.

Dayton, P. K., and Tegner, M. J. (1984). Catastrophic storms, El Niño, and patch stability in a southern California kelp community. *Science* 224, no. 4646, pp. 283–285.

Delavaud, C. C. (1985). Concerning the catastrophic floods in the north of Peru: "El Niño," myth and reality. *Revista Geografica* 101, pp. 133–139.

Dong Keqin, and Holland, G. J. (1994). A global view of the relationship between ENSO and tropical cyclone frequencies. *Acta Meteorologica Sinica* 8, no. 1, pp. 19–29.

Duffy, J. D. (March–April 2000). Effects of El Niño storms: California's Pacific Coast highway. *Transportation Research News* 207, pp. 21–22.

Egan, T. (September 26, 1997). El Niño's wrath: Floods and tickets. *New York Times*, p. A16.

El Niño events devastated two ancient civilizations. (March 3, 1990). *New Scientist* 125, p. 31.

Elston, R. (February 16, 1992). Rising fear of El Niño. *Los Angeles Times*, p. (B)1.

Ely, L. L., Enzel, Y., Baker, V. R., and Cayan, D. R. (October 15, 1993). A 5000-year record of extreme floods and climate change in the Southwestern United States. *Science* 262, pp. 410–412.

Evans, J. L., and Allan, R. J. (1992). El Niño/Southern Oscillation modification to the structure of the monsoon and tropical cyclone activity in the Australasian region. *International Journal of Climatology* 12, no. 6, pp. 611–623.

Fennell, T. (June 8, 1992). El Niño's angry year. *MacLean's* 105, p. 42.

Fennell, T., and Robinson, N. (March 9, 1998). The rage of El Niño. *Macleans* 111, pp. 34–35.

Goldenberg, S. B., and Shapiro, L. J. (1996). Physical-Mechanisms for the Association of El Niño and West-African Rainfall with Atlantic major hurricane activity. *Journal of Climate* 9, no. 6, pp. 1169–1187.

Gray, W. M. (1984). Atlantic seasonal hurricane frequency. I. El Niño and 30 mb quasibiennial Oscillation influences. *Monthly Weather Review* 112, no. 9, pp. 1649–1668.

Hadfield, P. (October 26, 1991). El Niño blamed for Japan's typhoons. *New Scientist* 132, p. 16.

Knowles, J. B., and Pielke, R. A. (1993). The Southern Oscillation and its effects on tornado activity in the United States. Unpublished manuscript.

Lander, M. A. (1994). An exploratory analysis of the relationship between tropical storm formation in the western North Pacific and ENSO. *Monthly Weather Review* 122, no. 4, pp. 636–651.

Motha, R., Puterbaugh, T., and Lundine, R. (1985). An ill wind: El Niño rainfall anomalies and regional crop yield variability (Latin America). *Mazingira* 8, no. 6, pp. 13–18.

Noel, J., and Changnon, D. (August 1998). A pilot study of winter cyclone frequency patterns associated with three ENSO parameters. *Journal of Climate*, pp. 2152–2159.

O'Brien, J. J., Richards, T. S., and Davis, A. C. (April 1996). The Effect of El-Niño on U.S. landfalling hurricanes. *Bulletin of the American Meteorological Society* 77, no. 4, pp. 773–774. http://www.coaps.fsu.edu/richards/paper.html

Oceanus 27, no. 2. (Summer 1984). [Entire issue is devoted to El Niño] Overall, an excellent summary of the phenomenon.

Pielke, Roger A., and Landsea, Christopher N. (1999). La Niña, El Niño, and Atlantic hurricane damage in the United States. *Bulletin of the American Meteorological Society* 80, no. 10, pp. 2027–2033.

Pisciottano, G., Diaz, A., Cazes, G., and Mechoso, C. R. (1994). El Niño–Southern Oscillation impact on rainfall in Uruguay. *Journal of Climate* 7, no. 8, pp. 1286–1302.

Rajeevan, M. (1989). Post monsoon tropical cyclone activity in the north Indian Ocean in relation to the El Niño/Southern Oscillation phenomenon. *Mausam* 40, no. 1, p. 43.

Researchers link midwest flooding to El Niño effect. (September 18, 1994). *Los Angeles Times* 18, p. A17.

Rhome, J. R., Niyogi, D. S., and Raman, S. (2000). Mesoclimatic analysis of severe weather and ENSO interactions in North Carolina. *Geophysical Research Letters* 27, no. 15, pp. 2269–2272.

Schonher, T., and Nicholson, S. E. (1989). The relationship between California rainfall and ENSO events. *Journal of Climate* 2, no. 11, p. 1258.

Stevens, W. K. (July 26, 1994). El Niño said to predict rain and crops in Africa. *The New York Times*, p. C4.

Stevens, W. K. (November 30, 1999). After the storm, an ecological bomb. *New York Times* on the Web.

Xavier, T. (May 2, 1995). Impact of ENSO episodes on autumn rainfall patterns near Sao Paulo, Brazil. *International Journal of Climatology* 15, pp. 571–584.

Zhu Bingyuan, and Li Donghang. (1992). The relationship between the El Niño events and the drought or excessive rain of northwest China during 1845 to 1988. *Scientia Atmospherica Sinica* 16, no. 2, pp. 185–192, 1992.

ENSO's Other Economic Impacts

Armstrong, L., and Veomett, E. (September 29, 1997). An ill wind for business: El Niño's mischief could set off a chain of economic calamities. *Business Week*, no. 3546, pp. 32–33.

Bell, A. (1986). El Niño, and prospects for drought prediction. *CSIRO Environmental Research* 49, pp. 12–18.

Bouncing back from disaster. (April 1984). *Science Digest*, no. 92, p. 16. A look at how nature slowly recovers after El Niño.

Changnon, David, Creech, Tamera, Marsili, Nathan, Murrel, William, and Saxinger, Michael. (June 1999). Interactions with a weather-sensitive decision-maker: a case study incorporating ENSO information into a strategy for purchasing natural gas. *Bulletin of the American Meteorological Society* 80, pp. 1117–1125.

Changnon, S. A. (1999). Impacts of the 1997–98 El Niño—generated weather in the United States. *Bulletin of the American Meteorological Society* 80, no. 9, pp. 1819–1827.

Childs, I.R.W., Hastings, P. A., and Auliciens, A. (1991). The acceptance of long-range weather forecasts: a question of perception? *Australian Meteorological Magazine* 39, no. 2, pp. 105–112.
Results from the survey, which was conducted prior to the release of ENSO-based long-range forecasts in Australia.

El Niño and U.S. Corn Belt rainfall. (May 27, 1992). *Weekly Weather and Crop Bulletin* 79, p. 10.

Flis, L. Using El Niño events in forecasting gas markets. *Rocky Mountain Oil Journal* 76, no. 4, pp. 1, 12.

Freeman, D. Gas price correlation? geologist charts El Niño effects. *American Association of Petroleum Geologists Explorer* 17, no. 8, pp. 32–33.

Glantz, M. H. (1984). Floods, fires, and famine: is El Niño to blame? *Oceanus* 27, no. 2, pp. 14–19.

Glantz, M. H. (1992). *Climate Variability, Climate Change and Fisheries* Cambridge: Cambridge University Press.
Discusses El Niño and variability in the northeastern Pacific salmon fishery.

Handler, P., and Handler, E. (1983). Climatic anomalies in the Tropical Pacific Ocean and corn yields in the United States. *Science* 220, no. 4602, pp. 1155–1156.

Association between El Niño events and annual corn production in the United States was investigated.

Hetter, K. (December 8, 1997). Outguessing El Niño: commodities markets feel the tension. *U.S. News & World Report* 123, pp. 87–88.

Hill, H.S.J. (2000). Comparing the value of Southern Oscillation Index-based climate forecast methods for Canadian and US wheat producers. *Agricultural and Forest Meteorology* 100, no. 4, pp. 261–272.

Iglesias, A., Erda, L. and Rosenzweig, C. (1996). Climate-change in Asia. a review of the vulnerability and adaptation of crop production. *Water, Air and Soil Pollution* 92, no. 1–2, pp. 13–27.

In South and Southeast Asia, there is concern about how climate change may affect El Niño/Southern Oscillation events, since these play a key role in determining agricultural production.

Keppenne, C. I. (June 1995) An ENSO signal in soybean futures prices. *Journal of Climate* 7, no. 10, pp. 1623–1627.

LeComte, D. (1984). A year of world wide extremes: flood and droughts (El Niño, 1983). *Weatherwise* 37, no. 1, pp. 8–11, 14–17, 19.

Meyers, S. D., and O'Brien, J. J. (1995). Variations in Mauna Loa carbon dioxide induced by ENSO. *EOS* 76, no. 52, http://www.coaps.fsu.edu/meyers/papers/CO2/eos.html

Miller, S. K. (July 23, 1994). Pacific winds of change predict African harvest. *New Scientist* 143, p. 10.

Namias, J., and Cayan, D. R. (1984). El Niño: implications for forecasting (USA). *Oceanus* 27, no. 2, pp. 41–47.

Niemira, Michael P. (February 1997). How does weather affect consumer behavior? Let me count the ways. *Chain Store Age*.

Orlove, B. S., Chiang, J. C., and Cane, M. A. (January 6, 2000). Forecasting Andean rainfall and crop yield from the influence of El Niño on Pleiades visibility. *Nature* 43, pp. 68–71.

Ponte, R. M. (1986). The statistics of extremes, with application to El Niño. *Reviews of Geophysics* 24, no. 2, pp. 285–297.

Stern, Paul C., and Easterling, William E., eds. (1994). *Making Climate Forecasts Matter*. University Corporation for Atmospheric Research.

Sterns, R. (December 1992). Salt water fishing: The El Niño effect. *Field and Stream* 97, pp. 110–113.

Tibbetts, J. (1996). Farming and fishing in the wake of El Niño. *Bioscience* 46, no. 8, pp. 566–569.

Weiss, C. (1985). The ENSO phenomenon: a new tool for predicting climatic change. *Ceres* 18, no. 6, pp. 36–38.

Wilson, Michael S. (March 1996). How El Niño affects the US natural gas industry. *Offshore Magazine* 56, no. 3, pp. 38–48.

Wilson, Michael S. (1997). The El Niño connection: correlation of El Niño events with profit cycles in the U.S. natural gas industry. *Geological Society of America Abstracts with Programs* 28, no. 7, p. A–360.

Yoshino, M., and Yasunari, T. (1986). Climatic anomalies of El Niño and anti-El Niño years and their socio-economic impacts in Japan. *Science Reports*—University of Tsukuba, Institute of Geoscience, Section A, vol. 7, pp. 41–53.

Zebiak, S. E., and M. A. Cane. (1987). A model El Niño/Southern Oscillation. *Monthly Weather Review* 115, pp. 2262–2278.

ENSO Health Impact References

Bouma, M. J., and Vanderkaay, H. J. (1994). Epidemic malaria in India and the El Niño Southern Oscillation. *The Lancet* 344, pp. 1638–1639.

Bouma, M. J., and Vanderkaay, H. J. (1996). The El Niño Southern Oscillation and the Historic malaria epidemics on The Indian subcontinent and Sri-Lanka. An early warning system for future epidemics. *Tropical Medicine & International Health* 1, no. 1, pp. 86–96.

Glantz, M. H. (1984). El Niño—should it take the blame for disasters? *Mazingira* 8, no. 1, pp. 21–26.

Glantz, M. H. (1984). Floods, fires, and famine: is El Niño to blame? *Oceanus* 27, no. 2, pp. 14–19.

Gueri, M., Gonzalez, C., and Morin, V. (1986). The effect of the floods caused by "El Niño" on health *Disasters* 10, no. 2, pp. 118–124.

Hales, S., Weinstein, P., and Woodward, A. (1996). Dengue fever epidemics in the South-Pacific driven by El Niño Southern Oscillation. *Lancet* 348, no. 9042 14, pp. 1664–1665.

Lindsay, S. W. (2000). Effects of 1997–1998 El Niño on highland malaria in Tanzania. *The Lancet* 355, no. 9208, pp. 989–990.

Moreira Cedeno, J. E. (1986). El Niño related health hazards. Rainfall and flooding in the Guayas river basin and its effects on the incidence of malaria 1982–1985. *Disasters* 10, no. 2, pp. 107–111.

Valencia Telleria, A. (1986). El Niño related health hazards. Health consequences of the floods in Bolivia in 1982. *Disasters* 10, no. 2, pp. 88–106.

Yang, X. B., and Scherm, Harald. (February 7, 1997). El Niño and infectious disease. *Science* 275, p. 739.

ENSO and the Environment

Adis, J., and Latif, M. (September 1996). Amazonian arthropods respond to El Niño. *Biotropica* 28, no. 3, pp. 403–407.

Anderson, D. J. (1989). Differential responses of boobies and other seabirds in the Galapagos to the 1986–87 El Niño-Southern Oscillation event. *Marine Ecology Progress Series* 52, no. 3, p. 209.

Anyamba, A. and Eastman, J. R. (1996). Interannual variability of NDVI over Africa and its relation to El Niño Southern Oscillation. *International Journal of Remote Sensing* 17, no. 13, pp. 2533–2548.

Armstrong, J. K., Williams, K., Huenneke, L. F., and Mooney, H. A.

(1998). Topographic position effects on growth depression of California Sierra Nevada pines during the 1982–83 El Niño. *Arctic & Alpine Research* 20, no. 3, pp. 352–357.

Arnold, C. (1998). *El Niño: Stormy Weather for People and Wildlife*. New York: Clarion Books.

Arntz, W. E., Pearcy, W. G., and Trillmich, F. (1991). Biological consequences of the 1982–83 El Niño in the Eastern Pacific. *Ecological Studies: Analysis and Synthesis* 88, pp. 22–44.

Arntz, W. E., and Tarazona, J. (1990). Effects of El Niño 1982–83 on benthos, fish and fisheries off the South American Pacific coast. In P. W. Glynn, ed., *Global Ecological Consequences of the 1982–83 El Niño-Southern Oscillation*, pp. 323–360.

Arntz, W. E., and Tarazona, J., Gallardo, V. A., Flores, L. A., and Salzwedel, H. (1991). Benthos communities in oxygen deficient shelf and upper slope areas of the Peruvian and Chilean Pacific coast, and changes caused by El Niño. In R. V. Tyson and T. H. Pearson, *Modern and Ancient Continental Shelf Anoxia*, pp. 131–154.

Aurioles, D., and Le Boeuf, B. J. (1991). Effects of the El Niño 1982–1983 on California sea lions in Mexico. *Ecological Studies: Analysis and Synthesis* 88, pp. 112–118.

Barber, R. T., and Chavez, F. P. (1983). Biological consequences of El Niño. *Science* 222, no. 4629, pp. 1203–1210.

Barber, R. T., and Kogelschatz, J. E. (1990), Nutrients and productivity during the 1982/83 El Niño. In P. W. Glynn, ed., *Global Ecological Consequences of the 1982–83 El Niño-Southern Oscillation*, pp. 21–53.

Barber, R. T., Sanderson, M. P., Lindley, S. T., Chai, F., Newton, J., Trees C. C., Foley, D. G., and Chavez, F. P. (1996). Primary productivity and its regulation in the Equatorial Pacific during and following the 1991–1992 El Niño. *Deep-Sea Research*, Part II—*Topical Studies* 43, no. 4–6, pp. 933–969.

Berkman, L. (June 22, 1992). El Niño brings natural disaster to sea creatures. *Los Angeles Times*, p. (A) 1.

Blanchot, J., Rodier, M., and Le Bouteiller, A. (1992). Effect of El Niño Southern Oscillation events on the distribution and abundance of phytoplankton in the Western Pacific tropical ocean along 165 degrees E. *Journal of Plankton Research*, 14, no. 1, pp. 137–156.

Boness, D. J., Oftedal, O. T., and Ono, K. A. (1991). The effect of El Niño on pup development in the California sea lion (*Zalophus californianus*). I. Early postnatal growth. *Ecological Studies Analysis and Synthesis* 88, pp. 173–179.

Bragg, J. (February 1994). NewsNet: Coho crash stuns the West Coast. *Pacific Fisherman* 15, pp. 16–17, 64–65.
El Niño's ocean warming causes havoc to fish and fishermen.

Brenner, J. (1999). *Southern Oscillation Anomalies and Their Relation to Florida Wildfires.* Florida Department of Agriculture Consumer Services.

Brodeur, R. D., Gadomski D. M., Pearcy, W. G., Batchelder, H. P., and Miller, C. B. (1985). Abundance and distribution of ichthyoplankton in the upwelling zone off Oregon during anomalous El Niño conditions. *Estuarine, Coastal & Shelf Science* 21, no. 3, pp. 365–378.

Campbell, L., Hongbin Liu, Nolla, H. A., and Vaulot, D. (1997) Annual variability of phytoplankton and bacteria in the subtropical North Pacific Ocean at Station ALOHA during the 1991–1994 ENSO event. *Deep-Sea Research* (Part I, vol. 44/2), pp. 167–192.

Carr, M. E., and Broad, K. (2000). Satellites, society, and the Peruvian fisheries during the 1997–98 El Niño. In D. Halpern, ed., *Satellites, Oceanography and Society. Elsevier Oceanography Series*, no. 63, pp. 171–191.

Carrasco, S., and Santander, H. (1987). The El Niño event and its influence on the zooplankton off Peru. *Journal of Geophysical Research* 92, Ch. 13, pp. 14, 405–14, 410.

Castilla, J. C., and Camus, P. A. (1992). The Humboldt—El Niño scenario: coastal benthic resources and anthropogenic influences, with particular reference to the 1982/83 ENSO. *South African Journal of Marine Science* 12, pp. 703–712.

Changnon, S. A. (1999). Impacts of the 1997–98 El Niño—Generated weather in the United States. *Bulletin of the American Meteorological Society* 80, no. 9, pp. 1819–1827.

Chavez, F. P. (1996). Forcing and biological impact of onset of the 1992 El Niño in central California. *Geophysical Research Letters* 23, no. 3., pp. 265–268.

Clark, L., Schreiber, R. W., and Schreiber, E. A. (1990). Pre- and post-El Niño Southern Oscillation comparison of nest sites for red-tailed tropicbirds breeding in the central Pacific Ocean. *The Condor* 92, no. 4, p. 886.

Codispoti, L. A., et al. (1986). High nitrite levels off northern Peru: a signal of instability in the marine denitrification rate. *Science* 233, no. 4769, pp 1200–1202.
Causes for the unusual conditions may include a cold anomaly that followed the 1982–1983 El Niño.

Coffroth, M. A., Lasker, H. R., and Oliver, J. K. (1990). Coral mortality outside of the eastern Pacific during 1982–1983: relationship to El Niño. In P. W. Glynn, ed., *Global Ecological Consequences of the 1982–83 El Niño-Southern Oscillation*, pp. 141–182.

Cole, J. E., Shen, G. T., Fairbanks, R. G., and Moore, M. (1993). Coral monitors of El Niño/Southern Oscillation dynamics across the equatorial Pacific. In H. F. Diaz and V. Markgraf, eds., *El Niño: Historical and Paleoclimatic Aspects of the Southern Oscillation*, pp. 349–375.

Colgan, M. W. (1990). El Niño and the history of eastern Pacific reef building. In P. W. Glynn, ed., *Global Ecological Consequences of the 1982–83 El Niño-Southern Oscillation*, pp. 183–232.

Costa, D. P., Antonelis, G. A., and DeLong, R. L. (1991). Effects of El Niño on the Foraging Energetics of the California Sea Lion. *Ecological Studies: Analysis and Synthesis* 88, pp. 156–165.

Cota, S. S, and Borrego, S. A. (1988). The "El Niño" effect on the phytoplankton of a north-western Baja California coastal lagoon. *Estuarine, Coastal & Shelf Science* 27 no. 1, pp. 109–115.

Couper-Johnston, R. (2000). *El Niño: The Weather Phenomenon That Changed the World*. London: Hodder & Stoughton.

Cowles, T. J., Barber, R. T., and Guillen, O. (1977). Biological consequences of the 1975 El Niño. *Science* 195, no. 4275, pp. 285–287.

Cruz, J. B., and Felipe, C. (1990), Effect of El Niño-Southern Oscillation conditions on nestling growth rate in the dark-rumped Petrel. *The Condor* 92, no. 1, p. 160.

Cucalon, E. (1987), Oceanographic variability off Ecuador associated with an El Niño event in 1982–3. *Journal of Geophysical Research* 92, no. C13, pp. 14309–14322.

Dandonneau, Y. (1986). Monitoring the sea surface chlorophyll concentration in the tropical Pacific: consequences of the 1982–83 El Niño. *Fishery Bulletin* 84, no. 3, pp. 687–695.

Dayton, L. (December 7, 1991). Forests and rivers dying in Australian drought. *New Scientist*, 132, p. 14.

Dayton, P. K., and Tegner, M. J. (1990). Bottoms beneath troubled waters: benthic impacts of the 1982–1984 El Niño in the temperate zone. In P. W. Glynn, ed., *Global Ecological Consequences of the 1982–83 El Niño-Southern Oscillation*, pp. 433–472.

DeLong, R. L., and Antonelis, G. A. (1991). Impact of the 1982–1983 El Niño on the northern fur seal population at San Miguel Island, California. *Ecological studies: analysis and synthesis* 88, pp. 75–83.

DeLong, R. L., Antonelis, G. A., Oliver, C. W., Stewart, B. S., Lowry, M. S., and Yochem, P. K. (1991). Effects of the 1982–1983 El Niño on several population parameters and diet of California sea lions on the California Channel Islands. *Ecological Studies: Analysis and Synthesis* 88, pp. 166–172.

Dillon, M. O., and Rundel, P. W. (1990). The botanical response of the Atacama and Peruvian Desert floras to the 1982–83 El Niño event. In P. W. Glynn, ed., *Global Ecological Consequences of the 1982–83 El Niño-Southern Oscillation*, pp. 487–504.

Doherty, T. J. (February 1994). El Niño threatens B.C. herring, salmon. *Pacific Fisherman* 15, pp. 17, 66.

Druffel, E.R.M., Dunbar, R. B., Wellington, G. M., and Minnis, S. A. (1990). Reef-building corals and identification of ENSO warming episodes. In P. W. Glynn, ed., *Global Ecological Consequences of the 1982–83 El Niño-Southern Oscillation*, pp. 233–253.

Dudley, N. J., and Hearn, A. B. (1993). El Niño effects hurt Namoi irrigated cotton growers, but they can do little to ease the pain. *Agricultural Systems* 42, no. 1–2, pp. 103–126.

Eakin, C. M. (1996). Where have all the carbonates gone: a model comparison of calcium-carbonate budgets before and after the 1982–1983 El Niño at Uva-Island in the eastern Pacific. *Coral Reefs* 15, no. 2, pp. 109–119.

Feingold, J. S. (1996). Coral survivors of the 1982–83 El Niño Southern-Oscillation, Galapagos Islands, Ecuador. *Coral Reefs* 15, no. 2, p. 108.

Feldkamp, S. D., DeLong, R. L., and Antonelis, G. A. (1991). Effects of El Niño 1983 on the foraging patterns of California sea lions (*Zalophus californianus*) near San Miguel Island, California. *Ecological Studies: Analysis and Synthesis*, 88, pp. 146–155.

Feldman, G., Clark, D., and Halpern, D. (1984). Satellite color observations of the phytoplankton distribution in the eastern Equatorial Pacific during the 1982–1983 El Niño (Galapagos Islands). *Science* 226, no. 4678, pp. 1069–1071.

Fiedler, P. C., Methot, R. D., and Hewitt, R. P. (1986). Effects of California El Niño 1982–1984 on the northern anchovy. *Journal of Marine Research* 44, no. 2, pp. 317–338.

Fraedrich, K. (1990). European Grosswetter during the warm and cold extremes of the El Niño/Southern Oscillation. *International Journal of Climatology* 10, no. 1, pp. 21–31.

Francis, J. M., and Heath, C. B. (1991). Population abundance, pup mortality, and copulation frequency in the California sea lion in relation to the 1983 El Niño on San Nicolas Island. *Ecological Studies: Analysis and Synthesis* 88, pp. 119–128.

Francis, J. M., and Heath, C. B. (1991). The effects of El Niño on the frequency and sex ratio of suckling yearlings in the California sea lion. *Ecological Studies: Analysis and Synthesis* 88, pp. 193–204.

Francou, B. (1985). "El Niño" and the drought in the High Central Andes of Peru and Bolivia. *Bulletin—Institut Francais des Etudes Andines* 14, no. 1–2, pp. 1–18.

Gentry, R. L. (1991). El Niño effects on adult northern fur seals at the Pribil of Islands. *Ecological Studies: Analysis and Synthesis* 88, pp. 84–93.

Gerard, V. A. (1984). Physiological effects of El Niño on giant kelp in southern California. *Marine Biology Letters* 5, no. 6, pp. 317–322.

Gibbs, H. L., and Grant, P. R. (1987). Ecological consequences of an exceptionally strong El Niño event on Darwin's finches. *Ecology* 68, no. 6, pp. 1735–1746.

Glantz, M. H. (1996). Currents of Change: El Niño and La Niña impacts on climate and society. Cambridge: Cambridge University Press. (Paperback.)

Glantz, M. H. (1992), Climate variability, climate change and fisheries. In *Climate Variability, Climate Change and Fisheries* Cambridge: Cambridge University Press). El Niño and variability in the northeastern Pacific salmon fishery.

Glantz, M. H. (1980). El Niño: lessons for coastal fisheries in Africa? *Oceanus* 23, no. 2, pp. 9–17.

Glynn, P. W. (1984). Widespread coral mortality and the 1982–83 El Niño warming event. *Environmental Conservation* 11, no. 2, pp. 133–146.

Glynn, P. W., and Colgan, M. W. (1992). Sporadic disturbances in fluctuating coral reef environments: El Niño and coral reef development in the eastern Pacific. *American Zoologist* 32, no. 6, pp. 707–718.

Glynn, P. W., and De Weerdt, W. H. (1991). Elimination of two reef-building hydrocorals following the 1982–83 El Niño warming event. *Science* 253, no. 5015, p. 69.

Glynn, P. W., Veron, J.E.N. and Wellington, G. M. (1996). Clipperton Atoll (Eastern Pacific) Oceanography, geomorphology, reef-building coral ecology and biogeography. *Coral Reefs* 15, no. 2, pp. 71–99. SST anomalies at Clipperton occur during Enso events and were greater at Clipperton in 1987 than during 1982–1983.

Grant, P. R., and Grant, B. R. (1987). The extraordinary El Niño event of 1982–83: effects on Darwin's finches on Isla Genovesa, Galapagos. *Oikos* 49, no. 1, pp. 55–66.

Guerra, C.G.C., and Portflitt, G. K. (1991). El Niño effects on Pinnipeds in northern Chile. *Ecological Studies: Analysis and Synthesis* 88, pp. 47–54.

Guinet, C., Jouventin, P., and Georges, J-Y. (1994). Long term population changes of fur seals *Arctocephalus gazella* and *Arctocephalus tropicalis* on subantarctic (Crozet) and subtropical (St. Paul and Amsterdam) Islands and their possible relationship to El Niño Southern Oscillation. *Antarctic Science* 6, no. 4, pp. 473–478.

Guzman, H. M., Cortes, J., Richmond, R. H., and Glynn, P. W. (1987). Effects of "El Niño-Southern oscillation" 1982/83 in the coral reefs at Isla del Cano, Costa Rica. *Revista de Biologia Tropical* 35, no. 2, pp. 325–332.

Hays, C. (1986). Effects of the 1982–83 El Niño on Humboldt penguin colonies in Peru. *Biological Conservation* 36, no. 2, pp. 169–180.

Heath, C. B., Ono, K. A., Boness, D. J., and Francis, J. M. (1991). The Influence of El Niño on female attendance patterns in the California Sea Lion. *Ecological Studies: Analysis and Synthesis* 88, pp. 138–145.

Huber, H. R. (1991). Changes in the distribution of California sea lions north of the breeding rookeries during the 1982–83 El Niño. *Ecological Studies: Analysis and Synthesis* 88, pp. 129–137.

Huber, H. R., Beckham, C., and Nisbet, J. (1991). Effects of the 1982–83 El Niño on northern elephant seals on the South Farallon Islands, California. *Ecological Studies: Analysis and Synthesis* 88, pp. 219–237.

Hughes, R. A. (1985). Notes on the effects of El Niño on the seabirds of the Mollendo district, southwest Peru in 1983. *Ibis* 127, no. 3, pp. 385–388.

Iglesias, A., Erda, L., and Rosenzweig, C. (1996). Climate-change in Asia: a review of the vulnerability and adaptation of crop production. *Water Air and Soil Pollution* 92, no. 1-2, pp. 13–27.
    In South and Southeast Asia, there is concern about how climate change may affect El Niño/Southern Oscillation events, since these play a key role in determining agricultural production.

Iverson, S. J., Oftedal, O. T., and Boness, D. J. The effect of El Niño on pup development in the California sea lion (*Zalophus californianus*). II. Milk Intake. *Ecological Studies: Analysis and Synthesis* 88, pp. 180–184.

Keeling, C. D., and Revelle, R. (1985). Effects of El Niño/Southern Oscillation on the atmospheric content of carbon dioxide. *Meteoritics* 20, no. 2, pt. 2, pp. 437–450.

Le Boeuf, B. J., and Reiter, J. (1991). Biological effects associated with El Niño, Southern Oscillation 1982–83, on northern elephant seals breeding at Ano Nuevo, California. *Ecological Studies: Analysis and Synthesis* 88, pp. 206–218.

Legler, D. M., Kelly, J. B., and O'Brien, J. J. (1997). Impact of ENSO-related climate anomalies on crop yields in the US. Submitted to *Climatic Change*, July 18, 1997.

Lough, J. M., and Fritts, H. C. (1990). Historical aspects of El Niño/Southern Oscillation—information from tree rings. In P. W., Glynn, ed., *Global Ecological Consequences of the 1982–83 El Niño-Southern Oscillation*, pp. 285–321.

Keppenne, C. L. (June 1995). An ENSO signal in soybean futures prices. *Journal of Climate* 1, pp. 1685–1689.

Kudela, C., and Chavez, F. P. (2000). Modeling the impact of the 1992 El Niño on new production in Monterey Bay, California. Deep-Sea Research. Part II. *Topical Studies in Oceanography* 47, no. 5/6, p. 1055–1076.

MacKenzie, D. (October 29, 1994). Missing salmon spawn political battle. *New Scientist* 144, p. 11.

Matarira, C. H. (1990). Drought over Zimbabwe in a regional and global context. *International Journal of Climatology* 10, no. 6, pp. 609–625.

Results show a significant correlation between the Southern Oscillation and seasonal rainfall over the region.

McKeon, G. M., and White, D. H. (1992). El Niño and better land management. *Search* 23, no. 6, pp. 197–200.

Mearns, A. J. (1988). The "Odd Fish": Unusual occurrences of marine life as indicators of changing ocean conditions. In D. F. Soule and G. S. Kleppel, eds., *Marine Organisms as Indicators.* New York: Springer-Verlag. Chapter 7, pp. 137–176.

Meserve, Peter., Yunger, John A., and Gutierrez, Julio R. (1995). Heterogeneous responses of small mammals to an El Niño Southern Oscillation event in northcenteral semiarid Chile and the importance of ecological scale. *Journal of Mammalogy* 76, pp. 580–595.

Millan-Nunez, E., and Gaxiola-Castro, G. (August 1, 1989). Spatial variability of phytoplankton in the Gulf of California during the El Niño 1983. *Nova Hedwigia* 49, no. 1/2, p. 113.

Miller, K. A., and Fluharty, D. L. (1992). El Niño and variability in the northeastern Pacific salmon fishery: implications for coping with climate change. In M. H. Glantz, ed., *Climate Variability, Climate Change and Fisheries,* pp. 49–88.

Miskelly, C. M. (1990). Effects of the 1982–83 El Niño event on two endemic landbirds on the Snares Islands, New Zealand. *The Emu* 90, p. 24.

Monastersky, R. (November 28, 1992). Do Antarctic seals feel El Niño? *Science News* 142, p. 382.

Moore, J. D. (1991). Cultural responses to environmental catastrophes: post-El Niño subsistence on the prehistoric north coast of Peru. *Latin American Antiquity* 2, no. 1, p. 27.

Moreira Cedeno, J. E. (1986). El Niño related health hazards. Rainfall and flooding in the Guayas river basin and its effects on the incidence of malaria 1982–1985. *Disasters* 10, no. 2, pp. 107–111.

Morliere, A., and Robert, J. P. (1986). Rainfall shortage and El Niño-Southern Oscillation in New Caledonia, southwestern Pacific. *Monthly Weather Review* 114, no. 6, pp. 1131–1137.

Motha, R., Puterbaugh, T., and Lundine, R. (1985). An ill wind: El Niño rainfall anomalies and regional crop yield variability (Latin America). *Mazingira* 8, no. 6, pp. 13–18.

Mullin, M. M. (1997). The demography of Calanus pacificus during winter–spring California El Niño conditions, 1992–1993: Implications for anchovy. *Fisheries Oceanography* 6, no. 1, pp. 10–18.

Murray, S. N., and Hurn, M. H. (1989). Variations in standing stocks of central California macrophytes from a rocky intertidal habitat before and during the 1982–1983 El Niño. *Marine Ecology Progress Series* 58, no. 1/2, p. 113.

Mysak, L. A. (1986). El Niño, interannual variability and fisheries in northeast Pacific Ocean. *Canadian Journal of Fisheries & Aquatic Sciences* 43, no. 2, pp. 464–497.

Nicholls, N., (1985). Impact of the Southern Oscillation on Australian crops. *Journal of Climatology* 5, no. 5, pp. 553–560.

Nicholls, N. (1991). The El Niño/Southern Oscillation and Australian vegetation. *Vegetatio* 91, no. 1/2, p. 23.

Ono, K. A., and Boness, D. J. (1991). The influence of El Niño on mother-pup behavior, pup ontogeny, and six ratios in the California sea lion. *Ecological Studies: Analysis and Synthesis*, 8, pp. 185–192.

Pearce, A. P., and Phillips, B. F. (1988). ENSO events, the Leeuwin Current, and larval recruitment of the western rock lobster. *Journal du Conseil—Conseil International pour l'Exploration de la Mer* 45, no. 1, pp. 13–21.

Pearcy, W. G., and Schoener, A. (1989). Changes in the marine biota coincident with the 1982–1983 El Niño in the northeastern subarctic Pacific Ocean. *Journal of Geophysical Research* 92, C13, pp. 14, 417–24, 428.

Pereyra Diaz, I., Angulo Cordova, Q., and Palma Grayeb, B. E. (1994). Effect of ENSO on the mid-summer drought in Veracruz State, Mexico. *Atmosfera* 7, no. 4, pp. 211–219.

Polonsky, A. B. (1994). Comparative study of the Pacific ENSO event of 1991–92 and the Atlantic ENSO-like event of 1991. *Australian Journal of Marine & Freshwater Research* 4, no. 4, pp. 705–725.

Ramos Duron, F., Gallegos Garcia, A., and de la Lanza Espino, G. (1986). Nutrient distributions in the shelf water off Guerrero, Mexico during the 1982–83 ENSO episode. *Revista Geofisica* 4, pp. 157–172.

Reid, F.M.H., and Steward, E. (1989). Nearshore microplanktonic assemblages off southern California in February 1983 during the El Niño event. *Continental Shelf Research* 9, no. 1, p. 37.

Reiter, E. R. (1978). *Journal of the Atmospheric Sciences* 35, no. 3, pp. 349–370. The interannual variability of the ocean–atmosphere system. The trade wind surges also are related to El Niño through a feedback involving the hydrological cycle and upwelling of cold water forced by Ekman pumping.

Ribic, C. A., Ainley, D. G., and Spear, L. B. (1992). Effects of El Niño and La Niña on seabird assemblages in the equatorial Pacific. *Marine Ecology Progress Series* 80, no. 2-3, pp. 109–124.

Richmond, R. H. (1990). The effects of the El Niño/Southern Oscillation on the dispersal of corals and other marine organisms, In P. W. Glynn, ed., *Global Ecological Consequences of the 1982–83 El Niño-Southern Oscillation*, pp. 127–140.

Robinson, G. R. (1987). Negative effects of the 1982–83 El Niño on Galapagos marine life. *Oceanus* 30, no. 2, pp. 42–48.

Rollins, H. B., Sendweiss, D. H., and Rollins, J. C. (1986). Effect of the

1982–1983 El Niño on bivalve mollusks. *National Geographic Research* 2, no. 1, pp. 106–112.

Sahley, C. T. (October 1996). Bay And hummingbird pollination of an autotetraploid columnar cactus, *Weberbauerocerous-Weberbaueri*. *American Journal of Botany* 83, no. 10, pp. 1329–1336.
Reduced bat pollination in 1993 is attributed to the reduced abundance of bats at the study site during a drought caused by El Niño.

Schereiber, R. W., Schreiber, E. A. (1984). Central Pacific seabirds and the El Niño Southern Oscillation: 1982 to 1983 perspectives. *Science* 225, no. 4663, pp. 713–716.

Shane, S. H. (July 20, 1995). Relationship between pilot whales and Rizzos dolphins at Santa Catalina Island, California, U.S.A. *Marine Ecology Progress Series* 123, pp. 5–11.

Sharp, G. D. (1993). Fishery catch records, El Niño/Southern Oscillation, and longer-term climate change as inferred from fish remains in marine sediments. In H. F. Diaz, and Maragraf, V. eds., *El Niño: Historical and Paleoclimatic Aspects of the Southern Oscillation*, pp. 379–417.

Smith, N. G. (1990). The Gulf of Panama and El Niño events: the fate of two refugee boobies from the 1982–83 event. In P. W. Glynn, ed., *Global Ecological Consequences of the 1982–83 El Niño-Southern Oscillation*, pp. 381–393.

Stewart, B. S., and Bochem, P. K. (1991). Northern elephant seals on the southern California Channel Island and El Niño. *Ecological Studies: Analysis and Synthesis* 88, pp. 234–246.

Tarazona, J., and Aritz, W. E. (1996). Impact of 2 El Niño events of different intensity on the hypoxic soft-bottom macrobenthos off the central Peruvian coast. *Marine Ecology-Pubblicazioni Della Stazione Zoologica Di Napoli* I, 17, no. 1–3, pp. 425–446.

Tarazona, J., Salzwedel, H., and Arntz, W. (1988). Oscillations of macrobenthos in shallow waters of the Peruvian central coast induced by El Niño 1982–83. *Journal of Marine Research* 46, no. 3, pp. 593–611.

Tegner, M. J., and Dayton, P. K. (1987). El Niño effects on southern California kelp forest communities, *Advances in Ecological Research* 17, pp. 243–279.

Thayer, Victoria G., and Barber, R. T. (October 1984). At sea with El Niño: the fish were gone, the birds were gone and the winds blew in the wrong direction. *Natural History* 93, pp. 4–12.

Tibetts, J. (1996). Farming and fishing in the wake of El Niño. *Bioscience* 46, no. 8, pp. 566–569.

Torres-Moye, G. and Alvarez-Borrego, S. (1989). Effects of the 1984 El Niño on the summer phytoplankton of a Baja California upwelling zone. *Journal of Geophysical Research* 92, Chapter 1, pp. 383–384, 386.

Trillmich, F. (1991). El Niño in the Galapagos Islands: a natural experiment. In H. A. Mooney, *Ecosystem Experiments* (New York: John Wiley & Sons), pp. 3–26.

Trillmich, F., and Dellinger, T. (1991). The effects of El Niño on Galapagos Pinnipeds. *Ecological Studies: Analysis and Synthesis* 88, pp. 66–74.

Trillmich, F., and Lunberger, D. (1985). Drastic efforts of El Niño on Galapagos pinnipeds. *Oecologia* 7, no. 1, pp. 19–22.

Valle, C. A., and Coulter, M. C. (1987). Present status of the flightless cormorant, Galapagos penguin and greater flamingo populations in the Galapagos Islands, Ecuador, after the 1982–83 El Niño. *Condor* 89, no. 2, pp. 276–281.

Valle, C. A., Cruz, F., Cruz, J. B., Merlen, G., and Coulter, M. C. (1987). The impact of the 1982–1983 El Niño—Southern Oscillation on seabirds in the Galapagos Islands, Ecuador. *Journal of Geophysical Research* 92, C13, pp. 14, 437–14, 444.

Vermeij, G. J. (1990). An ecological crisis in an evolutionary context: El Niño in the eastern Pacific, In P. W. Glynn, ed., *Global Ecological Consequences of the 1982–83 El Niño-Southern Oscillation*, pp. 505–517.

Vetter, R. E., Botosso, P. C. (1989). El Niño may affect growth behaviour of Amazonian trees. *GeoJournal* 19, no. 4, p. 419.

Wallace, M. P., and Temple, S. A. (1988). Impacts of the 1982–1983 El Niño on population dynamics of Andean condors in Peru. *Biotropica* 20, no. 2, pp. 144–150.

Wallace, J. M., and Vogel, S. (Spring 1994). *El Niño and Climate Prediction, Reports to the Nation on Our Changing Planet.* UCAR and NOAA Office of Global Programs.
  This report is available from UCAR Office of Interdisciplinary Earth Studies, P.O. Box 3000, Boulder, CO 80307–3000 (303) 497–2692.

Wikelski, M., and Thom, C. (2000). Marine iguanas shrink to survive El Niño. *Nature* 403, no. 6765, pp. 37–38.

Wolff, M. (1987). Population dynamics of the Peruvian scallop *Argopecten purpuratus* during the El Niño phenomenon of 1983. *Canadian Journal of Fisheries & Aquatic Sciences* 44, no. 10, pp. 1684–1691.

Woodhouse, C. A. (1993). Tree-growth response to ENSO events in the central Colorado Front Range. *Physical Geography* 14, no. 5, pp. 417–435.

Woodroffe, C. D., and Gagan, M. K. (2000). Coral microatolls from the Central Pacific record late Holocene El Niño. *Geophysical Research Letters* 27, no. 10, pp. 1511–1514.

Wootton, J. T., Power, M. E., Paine, R. T., and Pfister, C. A. (1996). Effects of productivity, consumers, competitors, and El Niño events on food-chain patterns in a rocky intertidal community. *Proceedings of the National Academy of Sciences of the United States of America* 93, no. 24–26, pp. 13855–13858.

Yakir, D., Levyadun, S., and Zangvil, A. (1996). El Niño and tree growth near Jerusalem over the last 20 years. *Global Change Biology* 2, no. 2, pp. 97–101.

Zabel, C. J., and Taggart, S. J. (1989). Shift in red fox, *Vulpes Vulpes,* mating system associated with El Niño in the Bering Sea. *Animal behavior* 38, no. 5, p. 830.

Zimmerman, R. C., and Robertson, D. L. (1985). Effects of El Niño on local hydrography and growth of the giant kelp. *Macrocystis pyrifera,* at Santa Catalina Island, California. *Limnology & Oceanography* 30, no. 6, pp. 1298–1302.

ENSO Prediction

Barnett, T. P. (1984). Prediction of the El Niño of 1982–3. *Monthly Weather Review* 112, no. 7, pp. 1403–1407.

Barnett, T. P., Bengtsson, L., Arpe, K., Flugel, M., Graham, N., Latif, M., Ritchie, J., Roeckner, E., Schlese, U., Schulzweida, U., and Tyree, M. (1994). Forecasting global ENSO-related climate anomalies. *Tellus, Series A (Dynamic Meteorology and Oceanography)*, 46A, no. 4, pp. 381–397.

Barnett, T. P., Latrif, M., Graham, N., Flugel, M., Pazan, S., and White, W. (1993). ENSO and ENSO-related Predictability. Part I: prediction of equatorial Pacific sea surface temperature with a hybrid coupled ocean–atmosphere model. *Journal of Climate* 6, no. 8, pp. 1545–1566.

Barnston, A. G., and Ropelewski, C. F. (1992). Prediction of ENSO episodes using canonical correlation analysis. *Journal of Climate* 5, no. 11, pp. 1316–1345.

Bell, A. (1986). El Niño and prospects for drought prediction. *CSIRO Environmental Research* 49, pp. 12–18.

Busalacchi, A. J., and O'Brien, J. J. An ocean model which predicts El Niño (Pacific Ocean). (1981). In J. P. McGreary, Jr., et al., eds., *Recent Progress in Equatorial Oceanography: A Report of the Final Meeting of Score Working Group* 47. Venice, pp. 75–86.

Cane, M. A., Zebiak, S. E., and Dolan, S. C. (1986). Experimental forecasts of El Niño. *Nature* 321, no. 6073, pp. 827–832.

Chen, D., Zebiak, S. E., Busalacchi, A. J., and Cane, M. A. (1995). An improved procedure for El Niño forecasting implications for predictability. *Science* 269, no. 5231, pp. 1699–1702.

Childs, I.R.W., Hastings, P. A., and Auliciens, A. (1991). The acceptance of long-range weather forecasts: a question of perception? *Australian Meteorological Magazine.* 39, no. 2, pp. 105–112.

Results from the survey, which was conducted prior to the release of ENSO-based long-range forecasts in Australia.

Davey, M. K. (1994). ENSO prediction experiments using a simple ocean-atmosphere model. *Tellus Series* A (Dynamic Meteorology and Oceanography) 46A, no. 4, pp. 465–480.

Davey, M. K., Anderson, D.L.T., and Lawrence, S. (1996). A simulation of variability of ENSO forecast skill. *Journal of Climate* 9, no. 1, pp. 240–246.

Fraedrich, K. (1988). El Niño/Southern Oscillation predictability. *Monthly Weather Review* 116, no. 5, pp. 1001–1012.

Gannon, R. (September 1986). Is this the key to long-range weather forecasting?: Solving the puzzle of El Niño. *Popular Science* 229, pp. 82–86, 118. New theories and measurements may enable predictions.

Glantz, M. H. (1994). Forecasting El Niño: science's gift to the 21st century. *Ecodecision* 12, pp. 78–81.

Hurlburt, H. E., Kindle, J. C., and O'Brien, J. J. (1976). A numerical simulation of the onset of El Niño. *Journal of Physical Oceanography* 6, no. 5, pp. 621–631.

Inoue, M. I., and O'Brien, J. J. A forecasting model for the onset of a major El Niño. *Monthly Weather Review* 112, no. 11, pp. 2326–2337.

Kruse, H. A., and Von Storch, H. (1986). A step towards long range weather prediction: the exceptional atmospheric circulation of January 1983 and its relation to El Niño. *Meteorologische Rundschau* 39, no. 5, pp. 152–160.

Penland, C., and Magorian, T. (1993). Prediction of Nino 3 sea surface temperatures using linear inverse modeling. *Journal of Climate* 6, no. 6, pp. 1067–1076.

Quinn, W. H. (1994). Monitoring and predicting El Niño invasions. *Journal of Applied Meteorology* 13, no. 7, pp. 825–830.

Sarachik, E. S. (1990). Predictability of ENSO. In M. E. Schlesinger, ed., *Climate–Ocean Interaction Proc. Workshop*. Oxford, 1988, pp. 161–171.

Spurgeon, D. (1984). The "Christ Child" and better climate prediction (El Niño). *Unesco Features*. 806, pp. 4–6.

Stern, P. C., and Easterling, W. E., eds. (1994). *Making Climate Forecasts Matter*. University Corporation for Atmospheric Research.

Stuller, J., (September 1986). Next El Niño: ocean warming is only one element in these complex and destructive weather patterns that may now be predictable. *Oceans* 19, pp. 18–23.

Swetnam, T. W., and Betancourt, J. L. (August 31, 1990). Fire-Southern Oscillation correlations in the southwestern United States. *Science* 249, pp. 1017–1020.

Tangang, F. T., Tang, B., Monahan, A. H., and Hseih, W. W. (1998). Forecasting ENSO events: a neural network–extended EOF approach. *Journal of Climate* 11, pp. 29–41.

Weiher, R. F. (2000). *Improving El Niño Forecasting: The Potential Economic Benefits.* Silver Spring, MD: National Oceanic and Atmospheric Administration, Central Library.

Weiss, C. (1985). The ENSO phenomenon: a new tool for predicting climatic change. *Ceres* 18, no. 6, pp. 36–38.

Wuethrich, B. (February 22, 1984). Predicting future climate. *EOS* 75, pp. 81–82. Looking at how far research has come and where it needs to go.

Zhang, Y., Wallace, J. M., and Battisti, D. S. (1997). ENSO-like interdecadal variability: 1900–93. *Journal of Climate* 10, pp. 1004–1020.

Zheng Da-Wei, Song Guo-xuan, and Luo Shi-fang. (1991). Prediction of El Niño events by the astronomical observation of the length of day. *Chinese Science Bulletin* 36, no. 4, pp. 305–308.

ENSO and the Oceans

Agenbag, J. J. (1996). Pacific ENSO events reflected in meteorological and oceanographic perturbations in the southern Benguela system. *South African Journal of Science* 92, no. 5, pp. 243–247.

Alexander, M. A. (1950). Simulation of the response of the North Pacific Ocean to the anomalous atmospheric circulation associated with El Niño. *Climate Dynamics* 5, no. 1, p. 53.

Allan, R. J., Beck, K., and Mitchell, W. M. (1990). Sea level and rainfall correlations in Australia: tropical links. *Journal of Climate* 3, no. 8, pp. 838–46.
Zero-lagged patterns show strong links with the El Niño/Southern Oscillation (ENSO) phenomenon over northern and eastern Australia.

Anderson, D.L.T., and McCreary, J. P. (1985). On the role of the Indian Ocean in a coupled ocean–atmosphere model of El Niño and the Southern Oscillation. *Journal of the Atmospheric Sciences* 42, no. 22, pp. 2439–2442.

Barnett, T. P. (1985). Variations in near-global sea level pressure. *Journal of Atmospheric Sciences* 42, pp. 478–501.

Barnett, T. P., Latrif, M., Graham, N., Flugel, M., Pazan, S., and White, W. (1993). ENSO and ENSO-related predictability. Part I: prediction of equatorial Pacific sea surface temperature with a hybrid coupled ocean–atmosphere model. *Journal of Climate* 6, no. 8, pp. 1545–1566.

Bigg, G. R., and Blundell, J. R. (1989). The equatorial Pacific Ocean prior to and during El Niño of 1982/83—a normal mode model view.

Quarterly Journal of the Royal Meteorological Society 115, no. 489, p. 1039.

Bigg, G. R., and Inoue, M. (1992). Rossby waves and El Niño during 1935–46. *Quarterly Journal of the Royal Meteorological Society* 118, no. 503, pp. 125–152.

Bin Li, and Clarke, A. J. (1994). An examination of some ENSO mechanisms using interannual sea level at the eastern and western equatorial boundaries and the zonally averaged equatorial wind. *Journal of Physical Oceanography* 24, no. 3, pp. 681–690.

Bjerknes, J. (1966). A possible response of the atmospheric Hadley circulation to equatorial anomalies of ocean temperatures. *Tellus* 18, pp. 820–829.

Blueford, J. (August 1988). El Niño: Ocean temperature affects weather and life cycles. *Instructor* 98, pp. 74–77.
A science teacher's approach to explaining El Niño.

Bongers, T., and Wyrtki, K. (1987). Sea level at Tahiti—a minimum of variability. *Journal of Physical Oceanography* 17, no. 1, pp. 164–168. Only the 1982–1983 El Niño event is clearly apparent in the data.

Brownlee, S. (January 1985). Death by degrees. *Discover* 6, pp. 44–48. Examining El Niño's widespread destruction of coral reefs.

Caldwell, P. (Summer 1992). Surfing the El Niño. *Mariners Weather Log* 36, pp. 60–64.

Cane, M. A. (1983). Oceanographic events during El Niño. *Science* 222, no. 4629, pp. 1189–1195.

Cane, M. A. (1984). Modeling sea level during El Niño. *Journal of Physical Oceanography* 14, no. 12, pp. 1864–1874.

Castilla, J. C., and Camus, P. A. (1992). The Humboldt–El Niño scenario: coastal benthic resources and anthropogenic influences, with particular reference to the 1982/83 ENSO. *South African Journal of Marine Science* 12, pp. 703–712.

Chao, Yi., Ghil, M., and McWilliams, J. C. (2000). Pacific interdecadal variability in this century's sea surface temperatures. *Geophysical Research Letters* 27, no. 15, pp. 2261–2264.

Chelton, D. B., Bernal, P. A., and Mcgowan, J. A. (1982). Large-scale interannual physical and biological interaction in the California Current (El Niño). *Journal of Marine Research* 40, no. 4, pp. 1095–1125.

Childers, D. L., Day J. W., Jr., and Muller, R. A. (1990). Relating climatological forcing to coastal water levels in Louisiana estuaries and the potential importance of El Niño-Southern Oscillation events. *Climate Research* 1 no. 1, pp. 31–42.

Chu, P. S., Frederick, J.; and Nash, A. J. (1991). Exploratory analysis of surface winds in the equatorial western Pacific and El Niño. *Journal of Climate* 4, no. 11, pp. 1087–102.

Clark, L., Schreiber, R. W., and Schreiber, E. A. (1990). Pre- and post-El Niño Southern Oscillation comparison of nest sites for red-tailed tropicbirds breeding in the central Pacific Ocean. *The Condor* 92, no. 4, p. 886.

Clarke, A. J. (1991). On the reflection and transmission of low-frequency energy at the irregular western Pacific Ocean boundary. *Journal of Geophysical Research* 96, suppl., pp. 3289–3305.

Clarke, A. J., and Van Gorder, S. (1994). On ENSO coastal currents and sea levels. *Journal of Physical Oceanography* 24, no. 3, pp. 661–680.

Colgan, M. W. (1990). El Niño and the history of eastern Pacific reef building. In P. W. Glynn, ed., *Global Ecological Consequences of the 1982–83 El Niño-Southern Oscillation*, pp. 183–232.

Congbin Fu, Diaz, H. F., and Fletcher, J. O. (1986). Characteristics of the response of sea surface temperature in the central Pacific associated with warm episodes of the Southern Oscillation. *Monthly Weather Review* 114, no. 9, pp. 1716–1738.

Cooke, R. (January 18, 1994). El Niño has a new found sister. *Newsday,* "Discovery," p. 57.

Discussion of La Niña, the cold phase of the Southern Oscillation.

Cucalon, E. (1987). Oceanographic variability off Ecuador associated with an El Niño event in 1982–3. *Journal of Geophysical Research* 92, no. C13, pp. 14309–14322.

Davey, M. K. (1994). ENSO prediction experiments using a simple ocean–atmosphere model. *Tellus, Series A (Dynamic Meteorology and Oceanography)* 46A, no. 4, pp. 465–480.

Dayton, P. K., and Tegner, M. J. (1990). Bottoms beneath troubled waters: benthic impacts of the 1982–1984 El Niño in the temperate zone: In P. W. Glynn, ed., *Global Ecological Consequences of the 1982–83 El Niño-Southern Oscillation*, pp. 433–472.

Delcroix, T., Boulanger, J.-P., Masia, F., and Menkes, C. (1994). Geosat-derived sea level and surface current anomalies in the equatorial Pacific during the 1986–1989 El Niño and La Niña. *Journal of Geophysical Research* 99, no. C12, pp. 25093–25107.

Delecluse, P., Servain, J., Levy, C., Arpe, K., and Bengtsson, L. (1994). On the connection between the 1984 Atlantic warm event and the 1982–1983 ENSO. *Tellus, Series A (Dynamic Meteorology and Oceanography)* 46A, no. 4, pp. 448–464.

Deser, C. (1993). Blackmon: Surface climate variations over the North Atlantic Ocean during winter: 1900–1989. *Journal of Climate* no. 6, pp. 1743–1753.

Dickson, R., Lazier, J., Meinke, J., Rhines, P., and Swift, J. (1996). Long-term coordinated changes in the convective activity of the North Atlantic. *Progress in Oceanography* 38, pp. 241–295.

Dickson, R. R., Meincke, J., Malmberg, S. A., and Lee, A. (1988). The "Great Salinity Anomaly" in the northern North Atlantic 1968–1982. *Oceanography* 20, pp. 103–151.

Donguy, J. R. (1987). Recent advances in the knowledge of the climatic variations in the tropical Pacific Ocean. *Progress in Oceanography* 19, no. 1, pp. 49–85.

An area particularly sensitive to interannual oscillations associated with El Niño.

Donguy, J. R., and Henin, C. (1978). Hydroclimatic anomalies in the South Pacific. *Oceanologica Acta* 1, no. 1, pp. 25–30.

Some six months after the appearance of the 'El Niño' phenomenon in the Eastern Pacific, a hydroclimatic anomaly is observed in the Western Pacific.

Donguy, J. R., and Henin, C. (1980). Climatic teleconnections in the western South Pacific with El Niño phenomenon. *Journal of Physical Oceanography* 10, no. 12, pp. 1952–1958.

Donguy, J. R., and Henin, C. (1980). Surface conditions in the eastern equatorial Pacific related to the intertropical convergence zone of the winds. *Deep-Sea Research, Part A (Oceanographic Research Papers)* 27, no. 9A, pp. 693–714.

During El Niño, the surface salinity is affected by the position of the intertropical convergence zone.

Donguy, J. R., Henin, C., Morliere, A., and Rebert, J. P. (1982). Thermal changes in the western tropical Pacific in relation to the wind field. *Deep-Sea Research, Part A (Oceanographic Research Papers)*, 29, no. 7A, pp. 869–892.

Alternate long periods of lifting and deepening of the thermocline appear in the western Pacific that may be roughly related to El Niño.

Druffel, E.R.M., Dunbar, R. B., Wellington, G. M., and Minnis, S. A. (1990). Reef-building corals and identification of ENSO warming episodes. In P. W. Glynn, ed., *Global Ecological Consequences of the 1982–83 El Niño-Southern Oscillation*, pp. 233–253.

Emery, W. J., and Hamilton, K. (1985). Atmospheric forcing of interannual variability in the northeast Pacific Ocean: connections with El Niño. *Journal of Geophysical Research* 90, no. C1, pp. 857–868.

Enfield, D. B. (1996). Relationship of inter-American rainfall to tropical Atlantic and Pacific SST variability. *Geophysical Research Letters* 23, no. 23, pp. 3305–3308.

Enfield, D. B., and Allen, J. S. (1980). On the structure and dynamics of monthly mean sea level anomalies along the Pacific Coast of North and South America. *Journal of Physical Oceanography* 10, no. 4, pp. 557–578.

The positive and negative sea-level anomalies, corresponding to El Niño–anti El Niño cycles, are well correlated throughout the tropics of both hemispheres and are detectable at the California stations.

Feldman, G., Clark, D., and Halpern, D. (1984). Satellite color observations of the phytoplankton distribution in the eastern Equatorial Pacific during the 1982–1983 El Niño (Galapagos Islands). *Science* 226, no. 4678, pp. 1069–1071.

Fennessy, M. J., and Shukla, J. (1991). Comparison of the impact of the 1982/83 and 1986/87 Pacific SST anomalies on time-mean predictions of atmospheric circulation. *Journal of Climate* 4, no. 4, pp. 407–423.

The primary aim of the study is to contrast the impact of the El Niño Pacific sea-surface temperature anomalies observed during the Northern Hemisphere winters of 1982–1983 and 1986–1987 on predictions with a global general circulation model.

Gill, A. E. (1982). Changes in thermal structure of the equatorial Pacific during the 1972 El Niño as revealed by bathythermograph observations. *Journal of Physical Oceanography* 12, no. 12, pp. 1373–1387.

Gill, A. E. (1983). An estimation of sea-level and surface-current anomalies during the 1972 El Niño and consequent thermal effects. *Journal of Physical Oceanography* 13, no. 4, pp. 586–606.

Glantz, M. H. (1992). Climate variability, climate change and fisheries. *Climate Variability, Climate Change and Fisheries*. Cambridge: Cambridge University Press.

Discusses El Niño and variability in the northeastern Pacific salmon fishery.

Gopinathan, C. K., and Sastry, J. S. (1990). Relationship between Indian summer monsoon rainfall and position of Pacific Ocean warm pool. *Indian Journal of Marine Sciences* 19, no. 4, pp. 246–250.

Gordon, H. B., and Hunt, B. G. (1991). Droughts, floods, and sea-surface temperature anomalies: a modelling approach. *International Journal of Climatology* 11, no. 4, pp. 347–365.

Three different sea-surface temperature anomaly patterns were explored, representative of both El Niño and anti-El Niño events.

Grove, J. S. (May–June 1984). At the heart of El Niño: Too warm waters surround the Galapagos Islands. *Oceans* 17, pp. 3–8.

Hansen D. V., and Bezdek, H. F. (1996). On the nature of decadal anomalies in North Atlantic Sea Surface Temperature. *Journal of Geophysical Research* 101, pp. 9749–9758.

Harrison, D. E., and Cane, M. A. (1984). Changes in the Pacific during the 1982–83 event (El Niño). *Oceanus* 27 no. 2, pp. 21–28.

Hayes, S. P., Mangum L. J., and Steffin, O. F. (July 1992). ATLAS buoy: watching for El Niño. *Sea Technology* 33, pp. 55–57.

Hisard, P. (1990). The Atlantic El Niño response revisited. *Hydrologie Continentale* 5, no. 2, pp. 87–104.

Hisard, P. (1980). The "El Niño" response of the Eastern Tropical Atlantic. *Oceanologica Acta* 3, no. 1, pp. 69–78.

Hsieh, W., Ware, D. M., and Thomson, R. E. (1995). Wind induced upwelling along the west coast of North America, 1899–1998. *Canadian Journal of Fisheries & Aquatic Sciences* 52, pp. 325–334.

Jacobs, G. A., Hurlburt, H. E., Kindle, J. C., Metzger, E. J., Mitchell, J. L., Teague, W. J., and Wallcraft, A. J. (1994). Decade-scale trans-Pacific propagation and warming effects of an El Niño anomaly. *Nature* 370, no. 6488, pp. 360–363.

Johnson, M. A., and O'Brien, J. J. (1990). The Northeast Pacific Ocean response to the 1982–3 El Niño. *Journal of Geophysical Research* 95, no. C5, pp. 7155–7166.

Joseph, P. V., Eischeid, J. K., and Pyle, R. J. (1994). Interannual variability of the onset of the Indian summer monsoon and its association with atmospheric features, El Niño, and sea surface temperature anomalies. *Journal of Climate* 7, no. 1, pp. 81–105.

Juillet-Leclerc, A., Labeyrie, L. D., Reyss, J. L., and Schrader, H. (1991). Temperature variability in the Gulf of California during the last century: a record of the recent strong El Niño. *Geophysical Research Letters* 18, no. 10, pp. 1889–1892.

Karl, D. M., Letelier, R., and Hebel, D. (1995). Ecosystem changes in the North Pacific subtropical gyre attributed to the 1991–92 El Niño. *Nature* 373, pp. 230–234.
Correlates ENSO, oceanic mixing, and the shift in mineral content of the Northern Pacific gyre.

Kawabe, M. (1993). Sea level variations due to equatorial Rossby waves associated with El Niño. *Journal of Physical Oceanography* 23, no. 8, pp. 1809–1822.

Kawabe, M. (1994). Mechanisms of interannual variations of equatorial sea level associated with El Niño. *Journal of Physical Oceanography* 24, no. 5, pp. 979–993.

Komar, Paul D., and Good, James W. (1989). Long term erosion impacts of the 1982–83 El Niño on the Oregon coast. In Coastal Zone '89: Proceedings of the sixth symposium on Coastal and Ocean Management held in Charleston, SC, July 11–14, 1989, pp. 3785–3794.
A study of one of the less expected ramifications of an El Niño event.

Kousky, V. E. 1997. (Spring 1997). Warm (El Niño) episode conditions return to the Tropical Pacific. *Mariners Weather Log* 41, no. 1, pp. 4–7.

Kousky, V. E., Kagano, M. T., and Cavalcanti, F. A. (1984). A review of the Southern Oscillation: oceanic-atmospheric circulation changes and related rainfall anomalies. *Tellus, Series A (Dynamic Meteorology and Oceanography)* 36A, no. 5, pp. 490–504.

Kudela, C., and Chavez, F. P. (2000). Modeling the impact of the 1992 El Niño on new production in Monterey Bay, California. Deep-sea research, Part II. *Topical Studies in Oceanography* 47, no. 5/6, pp. 1055–1076.

Kushnir, Y. (1994). Interdecadal variations in North Atlantic sea surface temperature and associated atmospheric conditions. *Journal of Climate* 7, pp. 141–157.

Lough, J. M. (1994). Climate variation and El Niño-Southern Oscillation events in the Great Barrier Reef: 1958 to 1987. *Coral Reefs* 13, no. 3, pp. 181–195.

Lukas, R., Hayes, S. P., and Wyrtki, K. (1984). Equatorial sea-level response during the 1982–83 El Niño. *Journal of Geophysical Research* 89, no. C6, pp. 10425–10430.

Lynn, R. J. (1983). The 1983–83 warm episode in the California Current. *Geophysical Research Letters* 10, no. 11, pp. 1093–1095.
The warm episode (El Niño) of 1982–1983 has produced elevated steric height (geopotential anomaly) along the coast of California and Baja California.

Markham, C. G., and McLain, D. R. (1977). Sea surface temperature related to rain in Ceara, northeastern Brazil. *Nature* 265, no. 5592, pp. 320–323.
The data suggest an association between below normal sea temperatures in the south Atlantic and the El Niño in the Pacific.

Martin, L., Flexor, J.-M., and Valentin, J.-L. (1988). The influence of the Pacific "El Niño" phenomenon on upwelling and on the climate of the region of Cabo Frio, on the Brazilian coast of the State of Rio de Janeiro. *Comptes Rendus—Academie des Sciences, Serie II*, 307, no. 9, pp. 1101–1105.

McCreary, J. P., Jr., and Anderson, D.L.T. (1984). A simple model of El Niño and the Southern Oscillation. *Monthly Weather Review* 112, no. 5, pp. 934–946.

Miller, K. A., and Fluharty, D. L. (1992). El Niño and variability in the northeastern Pacific salmon fishery: implications for coping with climate change. In M. H. Glantz, ed., *Climate Variability, Climate Change and Fisheries*, pp. 49–88.

Mo, K. C., and Kalnay, E. (1991). Impact of sea surface temperature anomalies on skill of monthly forecasts. *Monthly Weather Review* 119, no. 12, pp. 2771–2793.

Monastersky, R. (March 8, 1997). Pacific puts the brake on warming. *Science News* 151, p. 148.
A cooler eastern Pacific slows the rate of global warming.

Monastersky, R. (May 24, 1997). Pacific warmth augurs weird weather. *Science News* 151, p. 316.

Monastersky, R. (August 2, 1997). El Niño gathers steam in the Pacific. *Science News* 152, p. 75.
Surprisingly quick rise in ocean temperature on South America's western coast leads to prediction of recordbreaking temperatures and rainfall.

Moore, D., Hisard, P., McCreary, J., Merle, J., O'Brien, J., Picaut, J., Verstraete, J.-M., and Wunsch, C. (1978). Equatorial adjustment in the eastern Atlantic. *Geophysical Research Letters* 5, no. 8, pp. 637–640.
This explanation is analogous to current theories of El Niño in the Pacific Ocean.

Murakami, T. (1990). Equatorial disturbances, monsoons, and 1982–1983 ENSO. *Meteorology and Atmospheric Physics* 44, no. 1–4, pp. 85–100.

Murakami, Takio, and Sumathipala, W. L. (1989). Westerly bursts during the 1982/83 ENSO. *Journal of Climate* 2, no. 1, p. 71.

Nagai, T., Tokioka, T., Endoh, M., and Kitamura, Y. (1992). El Niño–Southern Oscillation simulated in an MRI atmosphere-ocean coupled general circulation model. *Journal of Climate* 5, no. 11, pp. 1202–1233.

Nicholls, N. (1984). The Southern Oscillation and Indonesian sea surface temperature. *Monthly Weather Review* 112, no. 3, pp. 424–432.

Nicholls, N. (1983). Predicting Indian monsoon rainfall from sea-surface temperature in the Indonesia-north Australia area. *Nature* 306, no. 5943, pp. 576–577,
Several studies have described associations between interannual variations of the Indian southwest monsoon and the ENSO phenomena, and methods for predicting monsoon rainfall based on these associations have been proposed.

Nouvelot, J. F., and Pourrut, P. (1984–1985). El Niño. Oceanic and atmospheric phenomenon. Importance 1982–1983 and its impact on the coastal region of Ecuador. *Cahiers—ORSTOM, Serie Hydrologie* 21, no. 1, pp. 39–65.

O'Brien, J. J. (Fall 1978). El Niño—An example of ocean/atmosphere interactions. *Oceanus* 21, pp. 40–46.

Paine, R. T. (1986). Benthic community-water column coupling during the 1982–1983 El Niño. Are community changes at high latitudes attributable to cause or coincidence? *Limnology & Oceanography* 31, no. 2, pp. 351–360.

Pariwono, J. I., Bye J.A.T., and Lennon, G.W. (1986). Long-period variations of sea-level in Australasia. *Geophysical Journal of the Royal Astronomical Society* 87, no. 1, pp. 43–54.
These features are related to the ENSO cycle which for the first time is linked, inter alia, with Southern Ocean mechanisms.

Patzert, W. C. (1996). El Niño expedition—a step toward ocean prediction. *Naval Research Reviews* 29, no. 4, pp. 18–21.

Penland, C., and Magorian, T. (1993). Prediction of Nino 3 Sea Surface Temperatures using linear inverse modeling. *Journal of Climate* 6, no. 6, pp. 1067–1076.

Petit, C. (November 11, 1994). Scientists say warming of Pacific may lead to another El Niño. *San Francisco Chronicle*, p. A8.

Philander, S.G.H. (1986). Unusual conditions in the tropical Atlantic Ocean in 1984. *Nature* 322, no. 6076, pp. 236–238.

Philander, S.G.H. (1989). El Niño and La Niña. A vast system of ocean–atmosphere exchanges covering the tropical Pacific. *American Scientist* 77, no. 5, p. 451.

Philander, S.G.H. (1989). El Niño, La Niña and the Southern Oscillation. San Diego: Academic Press, 294 pp.

Philander, S.G.H. (Summer 1992). El Niño. *Oceanus* 35, pp. 56–61.
An appraisal of El Niño and its relationship to tropical oceanography by one of the major researchers in the field.

Philander, S.G.H., and Delecluse, P. (1983). Coastal currents in low latitudes (with application to the Somali and El Niño currents). *Deep-Sea Research, Part A (Oceanographic Research Papers)* 30, no. 8A, pp. 887–902.

Philander, S.G.H., Hurlin, W.J., and Pacanowski, R. C. (1987). Initial conditions for a general circulation model of tropical oceans. *Journal of Physical Oceanography* 17, no. 1, pp. 147–57.
A general circulation model of the Tropical Pacific Ocean, which realistically stimulates El Niño of 1982–1985 has been used to determine how different initial conditions affect the model.

Picaut, J., and Delcroix, T. (1995). Equatorial wave sequence associated with warm pool displacements during the 1986–1989 El Niño–La Niña. *Journal of Geophysical Research* 100, no. C9, pp. 18393–18408.

Picaut, J., Ioualalen, M., Menkes, C., Delcroix, T., and McPhaden, M. J. (1996). Mechanism of the zonal displacements of the Pacific warm pool: implications for ENSO. *Science* 274, no. 5292, pp. 1486–1489.

Planet Earth 2: The Blue Planet. (1986). Produced by WQED Pittsburgh in association with the National Academy of Sciences. 60 min. Videocassette. Focuses on the oceans, including a section on El Niño.

Polonsky, A. B. (1994). Comparative study of the Pacific ENSO event of 1991–92 and the Atlantic ENSO-like event of 1991. *Australian Journal of Marine & Freshwater Research* 45, no. 4, pp. 705–725.

Prasad, K. D., and Singh, S. V. (1996). Seasonal variations of the Relationship between some Enso parameters and Indian rainfall. *International Journal of Climatology* 16, no. 8, pp. 923–933.

Quinn, W. H., and Neal, V. T. (1983). Long-term variations in the Southern Oscillation, El Niño, and Chilean subtropical rainfall. *Fishery Bulletin* 81, no. 2, pp. 363–374.

Rasmusson, E. M. (1984). El Niño: the ocean/atmosphere connection. *Oceanus* 27, no. 2, pp. 5–12.

Rasmusson, E. M., and Carpenter, T. H. (1983). The relationship between eastern equatorial Pacific sea surface temperatures and rainfall over India and Sri Lanka. *Monthly Weather Review* 111, no. 3, pp. 517–528. Monsoon season (June–September) precipitation data from 31 Indian subdivisions and mean monthly precipitation data from 35 Indian and Sri Lanka stations, spanning the period 1875–1979, were analyzed to determine the relationship between equatorial Pacific warm episodes (El Niño events) and interannual fluctuations in precipitation over India and Sri Lanka.

Rasmusson, E. M., and Carpenter, T. H. (1982). Variations in tropical sea surface temperature and surface wind fields associated with the Southern Oscillation/El Niño. *Monthly Weather Review* 110, no. 5, pp. 354–384.

Rasmusson, E. M., and Hall, J. M. (August 1983). El Niño: The great Equatorial Pacific Ocean warming event of 1982–1983. *Weatherwise* 36, pp. 166–175. A concise, readable report on this occurrence of El Niño.

Reiter, E. R. (1978). The interannual variability of the ocean-atmosphere system. *Journal of the Atmospheric Sciences* 35, no. 3, pp. 349–70. The trade wind surges also are related to El Niño through a feedback involving the hydrological cycle and upwelling of cold water forced by Ekman pumping.

Reyes, S., and Mejia-Trejo, A. (1991). Tropical perturbations in the eastern Pacific and the precipitation field over north-western Mexico in relation of the ENSO phenomenon. *International Journal of Climatology* 11, no. 5, pp. 515–528.

Rienecker, M. M., and Mooers, C.N.K. (1986). The 1982–3 El Niño signal off northern California: *Journal of Geophysical Research* 91, no. C5, pp. 6597–6608.

Roeckner, E. Oberhuber, J. M., Bacher, A., Christoph, M., and Kirchner, I. (1996). ENSO variability and atmospheric response in a global coupled atmosphere–ocean global climate model. *Climate Dynamics* 12, no. 11, pp. 737–754.

Rogers, J. C. (1997). North Atlantic storm track variability and its association to the North Atlantic Oscillation and climate variability of Northern Europe. *Journal of Climate* 10, no. 7, pp. 1635–1164.

Rusting, R. (October 1988). Pacific sea-saw; a natural feedback loop may explain El Niño's recurrences. *Scientific American* 259, pp. 20, 25. A good explanation of one of the theories of the mechanism of El Niño.

Sarachik, E. S. (1990). Predictability of ENSO: In M. E. Schlesinger, ed., *Climate–Ocean Interaction*. Proc. Workshop, Oxford, 1988, pp. 161–171.

Schell, I. I. (1965). The origin and possible prediction of the fluctuations in the Peru Current and upwelling. *Journal of Geophysical Research* 70, pp. 5529–5540.

Schmittner, A., Appenzeller, C., and Stocker, T. F. (2000). Enhanced Atlantic freshwater export during El Niño. *Geophysical Research Letters* 27, no. 8, pp. 1163–1167.

Strong, A. E. (1986). Monitoring El Niño using satellite based sea surface temperatures. *Ocean-Air Interactions* 1, no. 1, pp. 11–28.

Strub, P. T., James, C., Thomas, A.C., and Abbott, M.R. (1990). Seasonal and nonseasonal variability of satellite-derived surface pigment concentration in the California Current. *Journal of Geophysical Research* 95, no. C7, pp. 11501–11530.
Correlations between anomalous pigment concentrations and anomalous sea level heights are discussed.

Stuller, J. (September 1986). Next El Niño: ocean warming is only one element in these complex and destructive weather patterns that may now be predictable. *Oceans* 19, pp. 18–23.

Tang, T. Y., and Weisberg, R. H. (1984). On the equatorial Pacific response to the 1982/83 El Niño–Southern Oscillation event. *Journal of Marine Research* 42 no. 4, pp. 809–829.

Toole, J. M. (1984). Sea surface temperature in the equatorial Pacific, *Oceanus* 27, no. 2, pp. 29–34.

Tourre, Y. M., and White, W. B. (1995). ENSO signals in global upperocean temperature. *Journal of Physical Oceanography* 25, no. 6, pt. 1, pp. 1317–1332.

Vallis, G.K. (1986). El Niño: a chaotic dynamical system? *Science* 232, no. 447, pp. 243–245.

Volkov, Yu.N. (1980). The modelling of the El Niño phenomenon by an autooscillation process in the ocean–atmosphere system. *Izvestiya Akademii Nauk SSSR, Fizika Atmosfery i Okeana* 16, no. 11, pp. 1179–1180.

Wang, B. (1995). Transition from a cold to a warm state of the El Niño–Southern Oscillation cycle. *Meteorology and Atmospheric Physics* 56, no. 1–2, pp. 17–32.

Weare, B. C. (1982). El Niño and Tropical Pacific Ocean surface temperatures. *Journal of Physical Oceanography* 12, no. 1, pp. 17–27.

Weare, B. C. (1983). Interannual variation in net heating at the surface of the tropical Pacific Ocean. *Journal of Physical Oceanography* 13, no. 5, pp. 873–85.
Special emphasis is given to exploring the relationship between these variations and those in sea temperature associated with ENSO.

Weare, B. C. (1984). Interannual moisture variations near the surface of the tropical Pacific Ocean. *Quarterly Journal of the Royal Meteorological Society* 110, no. 464, pp. 489–504.
Special emphasis is given to understanding the relationship between variations in these quantities and ENSO.

Weare, B. C. (1987). Relationships between monthly precipitation and SST variations in the Tropical Pacific region. *Monthly Weather Review* 115, no. 11, pp. 2687–2698.

Wunsch, C. (1999). The interpretation of short climate records, with comments on the North Atlantic and Southern Oscillations. *Bulletin of the American Meteorological Society* 80, p. 257.

Wyrtki, K. (1975). El Niño—The dynamic response of the equatorial Pacific Ocean to atmospheric forcing. *Journal of Physical Oceanography* 5, no. 4, pp. 572–584.

Wyrtki, K. (1977). Advection in the Peru current as observed by satellite. *Journal of Geophysical Research* 82, no. 27, pp. 3939–3943.

Wyrtki, K. (1977). Sea level during the 1972 El Niño. *Journal of Physical Oceanography* 7, no. 6, pp. 779–787.

Wyrtki, K. (1979). The response of sea surface topography to the 1976 El Niño. *Journal of Physical Oceanography* 9, no. 6, pp. 1223–1231.

Wyrtki, K. (1982). The Southern Oscillation, ocean–atmosphere interaction and El Niño. *Marine Technology Society Journal* 16, no. 1, pp. 3–10.

Wyrtki, K. (1984). The slope of sea level along the equator during the 1982/1983 El Niño. *Journal of Geophysical Research* 89, no. C6, pp. 10419–10424.

Wyrtki, K. (1985). Sea level fluctuations in the Pacific during the 1982–83 El Niño. *Geophysical Research Letters* 12, no. 3, pp. 125–128.

Wyrtki, K. (1985). Water displacements in the Pacific and the genesis of El Niño cycles. *Journal of Geophysical Research* 90, no. C4, pp. 7129–7132.

Wyrtki, K., Stroup, E., Patzert, W., Williams, R., and Quinn, W. (1976). Predicting and observing El Niño. *Science* 191, no. 4225, pp. 343–346.

Wyrtki, K., and Wenzel, J. (1984). Possible gyre-gyre interaction in the Pacific Ocean. *Nature* 309, no. 5968, pp. 538–540.
Using sea-level data, the authors demonstrate that the subtropical gyres of the North and South Pacific oscillate with a period of about four years and that these oscillations are probably induced by the major El Niño events.

Yoo, J.-M., and Carton, J. A. (1990). Annual and interannual variation of the freshwater budget in the Tropical Atlantic Ocean and the Caribbean Sea. *Journal of Physical Oceanography* 20, no. 6, pp. 831–845.
Less rainfall in the region during the 11-year record occurred in the El Niño years of 1976–1977 and 1982–1983.

Zebiak, S. E. (1989). Oceanic heat content variability and El Niño cycles. *Journal of Physical Oceanography* 19, no. 4, p. 475.

Zeng Zhaomei, and Zhang Mingli. (1987). Relationship between the key region SST of the tropical eastern Pacific and air temperature of northeast China. *Scientia Atmospherica Sinica* 11, no. 4, pp. 382–389.

Cold summers in northeast China are consistent with the appearance of El Niño in the eastern equatorial Pacific during thirty years 1951–80.

Zhang X., Zhao H., and Ding Yihui. (1990). Sea level fluctuations in the Pacific during the 1986–7 El Niño. *Acta Meteorologica Sinica* 48, no. 4, pp. 424–432.

ENSO and the Global Atmosphere

Aceituno, P. (April 1993). El Niño, the Southern Oscillation and ENSO: confusing names for a complex ocean–atmosphere interaction. *Bulletin of the American Meteorological Society* 73, pp. 483–485.

A brief review of the history and nomenclature of El Niño.

Allan, R. J. (1988). El Niño Southern Oscillation influences in the Australasian region. *Progress in Physical Geography* 12, no. 3, pp. 313–348.

Allan, R. J., Lindesay, J., and Parker, D. (1986). El Niño Southern Oscillation and climatic variability. *El Niño Southern Oscillation and Climatic Variability*, 416 p. Perth, Australia: Csiro, 1996.

Allan, R. J., Nicholls, N., Jones, P. D., and Butterworth, I. J. (1991). A further extension of the Tahiti-Darwin SOI, early ENSO events and Darwin pressure. *Journal of Climate* 4, no. 7, pp. 743–749.

Allan, R. J., and Pariwono, J. I. (1990). Ocean-atmosphere interactions in low-latitude Australasia. *International Journal of Climatology* 10, no. 2, pp. 145–178.

Angell, J. K. (1988). Impact of El Niño on the delineation of tropospheric cooling due to volcanic eruptions. *Journal of Geophysical Research* 93, no. 4, pp. 3697–3704.

Angell, J. K. (1990). Variation in global tropospheric temperature after adjustment for the El Niño influence, 1958–89. *Geophysical Research Letters* 17, no. 8, pp. 1093–1096.

Angell, J. K. (1992). Evidence of a relation between El Niño and QBO, and for an El Niño in 1991–2. *Geophysical Research Letters* 19, no. 3, pp. 285–288.

Angell, J. K. and Korshover, J. (1984). Some long-term relations between equatorial sea-surface temperature, the four centers of action and 700 mb flow. *Journal of Climate and Applied Meteorology* 23, no. 9, pp. 1326–1332.

The only evidence of a precursor to warm equatorial SST (El Niño) is the relatively small distance between the centers of the Aleutian Low and Pacific High two seasons before warmest SST.

Angell, J. K., and Korshover, J. (1985). Displacements of the north circumpolar vortex during El Niño, 1963–83. *Monthly Weather Review* 113, no. 9, pp. 1627–1630.

Barnett, T. P. (1985). Variations in near-global sea level pressure. *Journal of Atmospheric Sciences* 42, pp. 478–501.

Barnston, A., and Livezey, R. E. (1987). Classification, seasonality and persistence of low-frequency atmospheric circulation patterns. *Monthly Weather Review* 115, pp. 1083–1126.

Barnston, A. G., and Livezey, R. (1989). A closer look at the effect of the 11 year solar cycle and QBO on Northern Hemispheric 700mb height and extratropical North American surface temperature. *Journal of Climate* 3, no. 11, pp. 1295–1313.

Barnston, A., and Livezey, R. (1991). Statistical prediction of the January–February mean northern hemisphere lower tropospheric climate from the 11 year solar cycle and the Southern Oscillation for west and east QBO phases. *Journal of Climate* 5, no. 2, pp. 249–262.

Barnston, A., Livezey, R., and Halpert, M. (1991). Modulation of Southern Oscillation—Northern Hemisphere mid-winter climate relationships by the QBO. *Journal of Climate* 4, no. 2, pp. 203–217.

Bjerknes, J. (1966). A possible response of the atmospheric Hadley Circulation to equatorial anomalies of ocean temperatures. *Tellus* 18, pp. 820–829.

Bjerknes, J. (1969). Atmospheric teleconnections from the equatorial Pacific. *Monthly Weather Review* 97, pp. 163–172.

Bjerknes, J. (1972). Large-scale atmospheric response to the 1964–65 Pacific equatorial warming. *Journal of Physical Oceanography* 2, pp. 212–217.

Blueford, J. (August 1988). El Niño: ocean temperature affects weather and life cycles. *Instructor* 98, pp. 74–77.

A science teacher's approach to explaining El Niño.

Bradley, R. S., Diaz, H. F., Kiladis, G. N., and Eischeid, J. K. (1987). ENSO signal in continental temperature and precipitation records. *Nature* 327, no. 6122, pp. 497–501.

Brahmananda Rao, V., and De Brito, J.I.B. (1985). Teleconnections between the rainfall over northeast Brazil and the winter circulation of Northern Hemisphere. *Pure and Applied Geophysics* 123, no. 6, pp. 951–959.

The occurrence of PNA pattern is interpreted as a connection between the Northern Hemisphere winter circulation and NE Brazil rainfall because of the ENSO phenomena.

Brock, R. C. (April 1984). El Niño and world climate: piecing together the puzzle. *Environment* 26, pp. 14–20, 37–39.

Bruman, C. (March 14, 1983). Genesis of a warm winter. *Maclean's 96*, p. 46. El Niño created unusual weather, and that created unusual problems.

Budin, G. R. (1985). International variability of Australian snowfall. *Australian Meteorological Magazine* 33, no. 3, pp. 145–159.
Exceptionally high snow-depth years are often followed by a low snow-depth year, which tend to occur during an El Niño.

Bunkers, M. J., Miller, J. R., Jr., and DeGaetano, A. T. (1996). An examination of El Niño–La Niña-related precipitation and temperature anomalies across the Northern Plains. *Journal of Climate* 9, no. 1, pp. 147–160.

Cayan, D. R. (1992). Latent and sensible heat flux anomalies over the northern oceans: the connection to monthly atmospheric circulation. *Journal of Climate* 5, pp. 354–369.

Cayan, D. R., and Webb, R. H. (1993). El Niño/Southern Oscillation and streamflow in the western United States: In H. F. Diaz and V. Markgraf, eds., *El Niño: Historical and Paleoclimatic Aspects of the Southern Oscillation*, pp. 29–68.

Chen, B., Smith, S. L., and Bromwich, D. H. (1996). Evolution of the tropospheric split jet over the South-Pacific Ocean during the 1986–89 Enso cycle. *Monthly Weather Review*, 124 no. 8, pp. 1711–1731.

Chen Lie-Ting. (1991). Relationship between the Northern Oscillation and monsoon rainfall in east China and air temperature in North America. *Chinese Science Bulletin* 36, no. 7, pp. 592–596.
This paper studies the influence of ENSO on Northern Hemisphere circulation, rainfall anomalies, and temperature.

Childers, D. L., Day, J. W., Jr, and Muller, R. A. (1990). Relating climatological forcing to coastal water levels in Louisiana estuaries and the potential importance of El Niño-Southern Oscillation events. *Climate Research* 1, no. 1, pp. 31–42.

Chiyu, Tik, and Chong, Lakseng. (1994). The effects of Asian monsoon and ENSO on Singapore rainfall—a preliminary study. *Journal of Tropical Meteorology/Redai Qixiang Xuebao* 10, no. 1, pp. 9–18.

Cohen, J., and Entekhabi, D. (1999). Eurasian snow cover variability and Northern Hemisphere climate predictability. *Geophysical Research Letters* 26, no. 3, pp. 345–348.

Cooke, R. (January 18, 1994). El Niño has a new found sister. *Newsday*, "Discovery," p. 57.
Discussion of La Niña, the cold phase of the Southern Oscillation.

Davydov, G. I., and Polonskii, A. B. (1996). Ocean–atmosphere variability in the Australo-Asiatic Region in connection with the El Niño Southern Oscillation. *Izvestiya Akademii Nauk Fizika Atmosfery I Okeana* 32, no. 3, pp. 383–396.

Delworth, T. L., Manabe, S., and Stouffer, R. J. (1993). Interdecadal variations of the thermohaline circulation in a coupled ocean–atmosphere model. *Journal of Climate* 6, pp. 1993–2011.

Deser, C. (1993). Blackmon: Surface climate variations over the North Atlantic Ocean during winter: 1900–1989: *Journal of Climate* 6, pp. 1743–1753.

Deser, C., and Wallace, J. M. (1990). Large-scale atmospheric circulation features of warm and cold episodes in the Tropical Pacific. *Journal of Climate* 3, no. 11, pp. 125–128.

DeWeaver, E., and Nigam, S. (1995). Influence of mountain ranges on the mid-latitude atmospheric response to El Niño events. *Nature* 378, no. 6558, pp. 706–708.

Donguy, J. R., and Henin, C. (1980). Climatic teleconnections in the western South Pacific with El Niño phenomenon. *Journal of Physical Oceanography* 10, no. 12, pp. 1952–1958.

Donguy, J. R., and Henin, C. (1980). Surface conditions in the eastern equatorial Pacific related to the intertropical convergence zone of the winds. *Deep-Sea Research, Part A (Oceanographic Research Papers)* 27, no. 9A, pp. 693–714.
During El Niño, the surface salinity is affected by the position of the intertropical convergence zone.

Estoque, M. A., Luque, J., Chandeck-Monteza, M., and Garcia, J. (1985). Effects of El Niño on Panama rainfall. *Geofisica Internacional* 24, no. 3, pp. 355–381.

Fennessy, M. J., and Shukla, J. (1991). Comparison of the impact of the 1982/83 and 1986/87 Pacific SST anomalies on time-mean predictions of atmospheric circulation. *Journal of Climate* 4, no. 4, pp. 407–23.
The primary aim of the study is to contrast the impact of the El Niño Pacific sea-surface temperature anomalies observed during the Northern Hemisphere winters of 1982–1983 and 1986–1987 on predictions with a global general circulation model.

Fraedrich, K. (1994). An ENSO impact on Europe? A review. *Tellus, Series A (Dynamic Meteorology and Oceanography)* 46A, no. 4, pp. 541–52.

Fraedrich, K., and Muller, K. (1992). Climate anomalies in Europe associated with ENSO extremes. *International Journal of Climatology* 12, no. 1, pp. 25–31.

Fraedrich, K., Muller, K., and Kuglin, R. (1992). Northern hemisphere circulation regimes during the extremes of the El Niño/Southern Oscillation. *Tellus, Series A (Dynamic Meteorology and Oceanography)* 44A, no. 1, pp. 33–40.

Gaffney, D. (1992). Seasonal climate summary southern hemisphere (autumn 1991): early indications of the formation of El Niño-type conditions. *Australian Meteorological Magazine* 40, no. 1, pp. 47–52.

Gan, M. A., and Rao, V. B. (1991). Surface cyclogenesis over South America. *Monthly Weather Review* 119, no. 5, pp. 1293–1302.
The occurrence of cyclogenesis is more frequent during the years of negative SOI (El Niño years) and less during the years of positive Southern Oscillation index.

Glantz, M. H., Katz, R., and Krenz, M., eds. (1987). Climate crisis: the societal impacts associated with the 1982–83 worldwide climate anomalies. Boulder, CO: National Center for Atmospheric Research.

Glantz, M. H., Katz, R. W., and Nicholls N., eds. (1991). Teleconnections linking worldwide climate anomalies. Cambridge: Cambridge University Press.

Goddard, L., and Philander, S. G. (2000). The energetics of El Niño and La Niña. *Journal of Climate* 13, no. 9, pp. 1496–1517.

Goldberg, R. A., Tisnado, M. G., and Scofield, R. A. (1987). Characteristics of extreme rainfall events in northwestern Peru during the 1982–3 El Niño period. *Journal of Geophysical Research* 92, no. C13, pp. 14225–14241.

Goodman, S. J., Buechler, D. E., and McCaul, E. W., Jr. The 1997–1998 El Niño event and related wintertime lightning variations in the southeastern United States. *Geophysical Research Letters* 27, no. 4, pp. 245–265.

Gordon, H. B., and Hunt, B. G. (1991). Droughts, floods, and sea-surface temperature anomalies: a modelling approach. *International Journal of Climatology* 11, no. 4, pp. 347–365.
Three different sea-surface temperature anomaly patterns were explored, representative of both El Niño and anti-El Niño events.

Graham, N. E., and White, W. B. (1988). The El Niño cycle: a natural oscillator of the Pacific Ocean–atmosphere system. *Science* 240, no. 4857, pp. 1293–1302.

Green, P. M., Legler, D. M., Miranda V, C. J. and O'Brien, J. J. The North American Climate Patterns Associated with the El Niño-Southern Oscillation. Available Online <http://www.coaps.fsu.edu/>

Guo Qiyun, and Wang Risheng. (1990). The relationship between the winter monsoon activity over east Asia and the El Niño events. *Acta Geographica Sinica* 45, no. 1, pp. 68–77.

Gutzler, D. S., and Rosen, R. D. (1992). Interannual variability of wintertime snow cover across the Northern Hemisphere. *Journal of Climate*, 5, pp. 1441–1447.

Hamilton, K. (1993). A general circulation model simulation of El Niño effects in the extratropical Northern Hemisphere stratosphere. *Geophysical Research Letters* 20, no. 17, pp. 1803–1806.

Juillet-Leclerc, A., Labeyrie, L. D., Reyss, J. L., and Schrader, H. (1991). Temperature variability in the Gulf of California during the last century: a record of the recent strong El Niño. *Geophysical Research Letters* 18, no. 10, pp. 1889–1892.

Kahya, E., and Dracup, J. A. (1994). The influences of Type 1 El Niño and La Niña events on streamflows in the Pacific Southwest of the United States. *Journal of Climate* 7, no. 6, pp. 965–976.

Kane, R. P. (1992). Relationship between QBOs of stratospheric winds, ENSO variability and other atmospheric parameters. *International Journal of Climatology* 12, no. 5, pp. 435–447.

Kane, R. P. (1999). Some characteristics and precipitation effects of the El Niño of 1997–1998. *Journal of Atmospheric and Solar-Terrestrial Physics* 61, no. 18, pp. 1325–1346.

Kane, R. P. (2000). El Niño/La Niña relationship with rainfall at Huancayo, in the Peruvian Andes. *International Journal of Climatology* 20, no. 1, pp. 63–72.

Keables, M. J. (1992). Spatial variability of mid-tropospheric circulation patterns and associated surface climate in the United States during ENSO winters. *Physical Geography* 13, no. 4, pp. 331–348.

Kok, C. J., and Opsteegh, J. D. (1985). Possible causes of anomalies in seasonal mean circulation patterns during the 1982–83 El Niño event. *Journal of the Atmospheric Sciences* 42, no. 7, pp. 677–694.

Kousky, V. E., Kagano, M. T., and Cavalcanti, F. A. (1984). A review of the Southern Oscillation: oceanic-atmospheric circulation changes and related rainfall anomalies. *Tellus, Series A (Dynamic Meteorology and Oceanography)* 36A, no. 5, pp. 490–504.

Kushnir, Y. (1994). Interdecadal variations in North Atlantic sea surface temperature and associated atmospheric conditions. *Journal of Climate* 7, pp. 141–157.

Labitzke, K., and Van Loon, H. (June 1989). Association between the 11 year solar cycle, the QBO and the atmosphere. Part III, Aspects of the association. *Journal of Climate*, pp. 554–565.

Li, Z., and Kafatos, M. (2000). Interannual variability of vegetation in the United States and its relation to El Niño/Southern Oscillation. *Remote Sensing of the Environment* 71, no. 3, pp. 239–247.

MacDonald, F. (2000). *El Niño*. Oxford: Oxford University Press.

Moses, T., Kiladis, G. N., Diaz, H. F., and Barry, R. G. (1987). Characteristics and frequency reversals in mean sea level pressure in the North Atlantic sector and their relationships to long-term temperature trends. *Journal of Climatology*, 7, pp. 13–30, 1987.

Nagai, T., Tokioka, T., Endoh, M., and Kitamura, Y. (1992). El Niño–Southern Oscillation simulated in an MRI atmosphere–ocean coupled general circulation model. *Journal of Climate* 5, no. 11, pp. 1202–1233.

Namias, J. (1976). Some statistical and synoptic characteristics associated with El Niño. *Journal of Physical Oceanography* 6, no. 2, pp. 130–138.

Nicholls, N. (1993). Historical El Niño/Southern Oscillation variability in the Australasian region. In H. F. Diaz and V. Markgraf, eds., *El Niño: Historical and Paleoclimatic Aspects of the Southern Oscillation*, pp. 151–173.

Nicholls, N., Lavery, B., Frederiksen, C., Drosdowsky, W., and Torok, S. (1996). Recent apparent changes in relationships between the El Niño–Southern Oscillation and Australian rainfall and temperature. *Geophysical Research Letters* 23, no. 23, pp. 3357–3360.

Nouvelot, J. F., and Pourrut, P. (1984–1985). El Niño. Oceanic and atmospheric phenomenon. Importance 1982–1983 and its impact on the coastal region of Ecuador *Cahiers—ORSTOM, Serie Hydrologie* 21, no. 1, pp. 39–65.

Nuñez, R. H., Richards, T. S., and O'Brien, J. J. (1996). Statistical analysis of Chilean precipitation anomalies associated with "El Niño Southern Oscillation" (1961–1994). http://www.coaps.fsu.edu/nunez/Paper1/paper1.html

O'Brien, J. J. (Fall 1978). El Niño—an example of ocean/atmosphere interactions. *Oceanus* 21, pp. 40–46.

Parker, D. E., and Folland, C. K. (1988). The nature of climatic variability. *Meteorology Magazine* 117, pp. 201–210.

Perlwitz, J., and Graf, H. F. (1995). The statistical connection between tropospheric and stratospheric circulation of the Northern Hemisphere in winter. *Journal of Climate* 8, pp. 2281–2295.

Philander, S.G.H. (1989). El Niño and La Niña. A vast system of ocean-atmosphere exchanges covering the Tropical Pacific. *American Scientist* 77, no. 5, p. 451.

Philander, S.G.H., Hurlin, W. J., and Pacanowski, R. C. (1987). Initial conditions for a general circulation model of tropical oceans. *Journal of Physical Oceanography* 17, no. 1, pp. 147–157.
A general circulation model of the Tropical Pacific Ocean, which realistically stimulates El Niño of 1982–1983, has been used to determine how different initial conditions affect the model.

Prasad, K. D., and Singh, S. V. (1996). Seasonal variations of the relationship between some Enso parameters and Indian rainfall. *International Journal of Climatology* 16, no. 8, pp. 923–933.

Qian Weihong. (1997). The understanding of ENSO cycle mechanism and ENSO potential prediction ability. *Acta Meteorologica Sinica* 11/1, pp. 105–118.

Ramage, C. S., and Hori, A. M. (1981). Meteorological aspects of El Niño. *Monthly Weather Review* 109, no. 9, pp. 1827–1835.

Rasmusson, E. M. (1984). El Niño: the ocean/atmosphere connection. *Oceanus* 27, no. 2, pp. 5–12.

Rasmusson, E. M., and Carpenter, T. H. (1982) Variations in tropical sea surface temperature and surface wind fields associated with the Southern Oscillation/El Niño. *Monthly Weather Review* 110, no. 5, pp. 354–384.

Rasmusson, E. M., and Carpenter, T. H. (1983). The relationship between eastern equatorial Pacific sea surface temperatures and rainfall over India and Sri Lanka. *Monthly Weather Review* 111, no. 3, pp. 517–528.
Monsoon season (June–September) precipitation data from 31 Indian subdivisions and mean monthly precipitation data from 35 Indian and Sri Lanka stations, spanning the period 1875–1979, were analyzed to determine the relationship between equatorial Pacific warm episodes (El Niño events) and interannual fluctuations in precipitation over India and Sri Lanka.

Rasmusson, E. M., and Wallace, J. M. (1983). Meteorological aspects of the El Niño/Southern Oscillation. *Science* 222, no. 4629, pp. 1195–1202.

Reiter, E. R. (1978). The interannual variability of the ocean–atmosphere system. *Journal of the Atmospheric Sciences* 35, no. 3, pp. 349–70.
The trade wind surges also are related to El Niño through a feedback involving the hydrological cycle and upwelling of cold water forced by Ekman pumping.

Renwick, J. A. and Wallace, J. M. (1996). Relationships between North Pacific wintertime blocking, El Niño, and the PNA pattern. *Monthly Weather Review* 124, no. 9, pp. 2071–2076.

Reyes, S., and Mejia-Trejo, A. (1991). Tropical perturbations in the eastern Pacific and the precipitation field over north-western Mexico in relation of the ENSO phenomenon. *International Journal of Climatology* 11, no. 5, pp. 515–528.

Roeckner, E., Oberhuber, J. M., Bacher, A., Christoph, M., and Kirchner, I. (1996). ENSO variability and atmospheric response in a global coupled atmosphere–ocean global climate model. *Climate Dynamics* 12, no. 11, pp. 737–754.

Rogers. J. C. (1990). Patterns of low-frequency monthly sea level pressure variability (1899–1986) and associated wave cyclone frequencies. *Journal of Climate* 3, pp. 1364–1379.

Rogers, J. C. (1997). North Atlantic storm track variability and its association to the North Atlantic Oscillation and climate variability of Northern Europe. *Journal of Climate* 10, no. 7, pp. 1635–1164.

Rong-Hua Zhang, and Endoh, M. (1994). Simulation of the 1986–1987 El Niño and 1988 La Niña events with a free surface Tropical Pacific Ocean general circulation model. *Journal of Geophysical Research* 99, no. C4, pp. 7743–7759.

Ronghui, Huang, and Yifang, Wu. (1989). The influence of ENSO on the summer climate change in China and its mechanism. *Advances in Atmospheric Sciences* 6, no. 1, p. 21.

Ropelewski, C. F., and Halpert, M. S. (1986). North American precipitation and temperature patterns associated with the El Niño/Southern Oscillation (ENSO). *Monthly Weather Review* 114, no. 12, pp. 2352–2362.

Ropelewski, C. F., and Halpert, M. S. (1987). Global and regional scale precipitation patterns associated with the El Niño/Southern Oscillation. *Monthly Weather Review* 115, no. 8, pp. 1606–1626.

Sasi, M. N. (1994). A relationship between equatorial lower stratospheric QBO and El Niño. *Journal of Atmospheric and Terrestrial Physics* 56, no. 12, pp. 1563–1570.

Serreze, M., Carse, F., and Barry, R. G. (1997). Icelandic low cyclone activity: climatological features, linkages with the NAO, and relationship with recent changes in the Northern Hemisphere circulation. *Journal of Climate* 10, pp. 453–464.

Shabbar, A., and Khandekar, M. (1966). The impact of El Niño–Southern Oscillation on the temperature-field over Canada. *Atmosphere–Ocean* 34, no. 2, pp. 401–416.

Stockton, C. W., and Glueck, M. F. (1999). Long-term variability of the North Atlantic oscillation (NAO). Proceedings of the American Meteorological Society Tenth Symposium on Global Change Studies, January 1999, Dallas, TX, pp. 290–293.

Tai, K. C. (1987). Establishing qualitative hypotheses on the influence of El Niño on Metropolitan Adelaide water supply (Australia). *Water International* 12, no. 1, pp. 8–14.

Tapley, T. D., and Waylen, Peter R. (1990). Spatial variability of annual precipitation and ENSO events in western Peru. *Hydrological Sciences Journal* 35, no. 4, p. 429.

Tomita, T., and Yasunari, T. (1996). Role of the northeast winter monsoon on the Biennial oscillation of the ENSO/monsoon system. *Journal of the Meteorological Society of Japan* 74, no. 4, pp. 399–413.

Trenberth, K. E. (1990). Recent observed interdecadal climate changes in the Northern Hemisphere. *Bulletin of the American Meteorological Society* 71, no. 7, pp. 988–993.
North Pacific changes appear to be linked through teleconnections to tropical atmosphere–ocean interactions and the frequency of El Niño.

Trenberth, K. E. (1991). *General Characteristics of the El Niño Southern Oscillation Teleconnections Linking Worldwide Climate Anomalies.* Cambridge: Cambridge University Press.

Vallis, G. K. (1986). El Niño: a chaotic dynamical system? *Science* 232, no. 447, pp. 243–245.

Van Loon, H. (September 1988). Association between the 11 year solar cycle, the QBO and the atmosphere. Part II: Surface and 700 mb in the Northern Hemisphere winter. *Journal of Climate*, pp. 905–920.

Van Loon, H., and Rogers, J. C. (1978). The seesaw in winter temperatures between Greenland and Northern Europe. Part I: General description. *Monthly Weather Review* 106, pp. 296–310.

Van Loon, H., and Shea, D. J. (1985). The Southern Oscillation. Part IV. The precursors south of 15 degrees S to the extremes of the Oscillation. *Monthly Weather Review* 113, no. 12, pp. 2063–2074.

Van Loon, H., and Williams, J. (1976). The connection between trends of mean temperature and circulation at the surface. Part I: Winter. *Monthly Weather Review* 104, pp. 365–380.

Van Oldenborgh, G. J., Burgers, G., and Tank, A. K. (2000). On the El Niño teleconnection to spring precipitation in Europe. *International Journal of Climatology* 20, no. 5, pp. 565–574.

Venne, D., and Dartt, D. (February 1990). An examination of possible solar cycle/QBO effects on the Northern Hemisphere troposphere. *Journal of Climate*, pp. 272–281.

Verma, R. K. (1990). Recent monsoon variability in the global climate perspective. *Mausam* 41, no. 2, pp. 315–320.

Von Storch, H., and Kruse, H. A. (1985). The extra-tropical atmospheric response to El Niño event—a multivariate significance analysis. *Tellus, Series A (Dynamic Meteorology and Oceanography)* 37A, no. 4, pp. 361–377.

Wallace, J. M., Zhang, Y., and Renwick, J. A. (1995). Dynamic contribution to hemispheric mean temperature trends. *Science* 270, pp. 780–783.

Wang Shaowu, Ma Liang, Chen Zhenhua, and Zhang Qiwen. (1986). The western and central Pacific rainfall and the El Niño events, *Acta Meteorologica Sinica* 44, no. 4, pp. 403–410.

Wang, W. C., and Li, Kerang. (1990). Precipitation fluctuation over semi-arid region in northern China and the relationship with El Niño/Southern Oscillation. *Journal of Climate* 3, no. 7, p. 769.

Waylen, P. R., and Caviedes, C. N. (1987). El Niño and annual floods in coastal Peru: In L. Mayer and D. Nash, eds., *Catastrophic Flooding*, pp. 57–77.

Waylen, P. R., and Caviedes, C. N. (1980). El Niño and annual floods on the north Peruvian littoral. *Journal of Hydrology*, 89, no. 1–2, pp. 141–156.

Waylen, P. R., Quesada, M. E., and Caviedes, C. N. (1994). The effects of El Niño–Southern Oscillation on precipitation in San Jose, Costa Rica. *International Journal of Chronology* 14, no. 5, pp. 559–568.

Weber, G.-R. (1990). North Pacific circulation anomalies, El Niño and anomalous warmth over the North American continent in 1986–1988: possible causes of the 1988 North American drought. *International Journal of Climatology* 10, no. 3, pp. 279–289.

Webster, F. (1984). Studying El Niño on a global scale (TOGA). *Oceanus* 27 no. 2, pp. 58–62.

Webster, P. J., and Yang, S. (1992). Monsoon and ENSO: selectively interactive systems. *Quarterly Journal of the Royal Meteorological Society* 118, no. 507, pp. 877–926.

Whetton, P. H., and Rutherfurd, I. (1994). Historical ENSO teleconnections in the eastern hemisphere. *Climatic Change* 28, no. 3, pp. 221–253.

Worldwide Climate Anomalies. (New York): Cambridge University Press. A college-level text that is written clearly, approaching the issues of highly dispersed effects of El Niño.

Wunsch, C. (1999). The interpretation of short climate records, with comments on the North Atlantic and Southern Oscillations. *Bulletin of the American Meteorological Society* 80, p. 257.

Xinglin, T. (1990). A diagnostic analysis of winter: atmospheric circulation during the 1982–1983 ENSO event. *Advances in Atmospheric Sciences* 7, no. 1, p. 57.

Zeng Zhaomei, and Zhang Mingli. (1987). Relationship between the key region SST of the Tropical Eastern Pacific and air temperature of northeast China. *Scientia Atmospherica Sinica* 11, no. 4, pp. 382–389.
Cold summers in northeast China are consistent with the appearance of El Niño in the eastern equatorial Pacific during thirty years 1951–1980.

Zhang Jijia, Lia Yueqing, Lei Zhaochong and Sun Zhaobo. (1992). Characteristics of the Northern Hemisphere 500hPa diabatic heat flow field in years of El Niño and anti-El Niño, *Acta Meteorologica Sinica* (English Edition) 6, no. 2, pp. 159–169.

Historical ENSO Events

Acharya, A. (July/August 1995). El Niño: fingerprint of climate change? *World Watch* 8, p. 6.

Allan, R. J., Nicholls, N., Jones, P. D., and Butterworth, I. J. (1991). A further extension of the Tahiti–Darwin SOI, early ENSO events and Darwin pressure. *Journal of Climate* 4, no. 7, pp. 743–749.

Anderson, R. Y., Linsley, B. K., and Gardner, J. V. (1990). Expression of seasonal and ENSO forcing in climatic variability at lower than ENSO frequencies: evidence from Pleistocene marine varves off California. *Palaeogeography, Palaeoclimatology, Palaeoecology.* 78, no. 3/4, p. 287.

Anderson, R. Y., (1993). Long-term changes in the frequency of occurrence of El Niño events. In H. F. Diaz and V. Markgraf, eds., *El Niño: Historical and Paleoclimatic Aspects of the Southern Oscillation*, pp. 193–200.

Anderson, R. Y., Soutar, A., and Johnson, T. C. (1993). Long-term changes in El Niño/Southern Oscillation: evidence from marine and lacustrine sediments: In H. F. Diaz and V. Markgraf, eds., *El Niño: Historical and Paleoclimatic Aspects of the Southern Oscillation*, pp. 419–433.

Anderson, W. L., Robertson, D. M., and Magnuson, J. J. (1996). Evidence of recent warming and El Niño-related variations in ice breakup of Wisconsin lakes. *Limnology and Oceanography* 41, no. 5, pp. 815–821.

Angell, J. K., and Korshover, J. (1984). Some long-term relations between equatorial sea-surface temperature, the four centers of action and 700 mb flow. *Journal of Climate and Applied Meteorology* 23, no. 9, pp. 1326–1332.
The only evidence of a precursor to warm equatorial SST (El Niño) is the relatively small distance between the centers of the Aleutian Low and Pacific High two seasons before warmest SST.

Angell, J. K., and Korshover, J. (1985). Displacements of the north circumpolar vortex during El Niño, 1963–83. *Monthly Weather Review* 113, no. 9, pp. 1627–1630.

Anyamba, A. and Eastman, J. R. (1996). Interannual variability of NDVI over Africa and its relation to El Niño Southern Oscillation. *International Journal of Remote Sensing* 17, no. 13, pp. 2533–2548.

Barricklow, D. (November 1992). Unlocking the mysteries of El Niño. *Choices*, pp. 32–35. Scientists seek new ways of tracking these critical events.

Begley, S. (October 6, 1997). Searching for El Niño. *Newsweek* 130, pp. 54–57.

Bergeron, L. (1996). Will El Niño become El Hombre? *New Scientist* 149, p. 15.

Bergman, K. H. (1987). The global climate of September–November 1986: a moderate ENSO warming develops in the Tropical Pacific. *Monthly Weather Review* 115, no. 10, pp. 2524–2541.

Bjerknes, J. (1972). Large-scale atmospheric response to the 1964–65 Pacific equatorial warming. *Journal of Physical Oceanography* 2, pp. 212–217.

Bongers, T., and Wyrtki, K. (1987). Sea level at Tahiti—a minimum of variability. *Journal of Physical Oceanography* 17, no. 1, pp. 164–168. Only the 1982–1983 El Niño event is clearly apparent in the data.

Bradley, R. S., Diaz, H. F., Kiladis, G. N., and Eischeid, J. K. (1987). ENSO signal in continental temperature and precipitation records. *Nature* 327, no. 6122, pp. 497–501.

Brownlee, S. (January 1985). Death by degrees. *Discover* 6, pp. 44–48. Examining El Niño's widespread destruction of coral reefs.

Bruman, C. (March 14, 1983). Genesis of a warm winter. *Maclean's* 96 p. 46. El Niño created unusual weather and that created unusual problems.

Burnham, L. (March 1989). The summer of 1988: a closer look at last year's drought. *Scientific American*, p. 21.

Busalacchi, A. J., Takeuchi, K., and O'Brien, J. J. (1983). Interannual variability of the equatorial Pacific—revisited. *Journal of Geophysical Research* 88, no. C12, pp. 7551–7562.
The additional eight years include the 1972 and 1976 Los Niños and the aborted event of 1975.

Caviedes, C. N. (1984). El Niño 1982–83. *Geographical Review* 74, no. 3, pp. 267–290.

Caviedes, C. N. (1984). Geography and the lessons from El Niño. *Professional Geographer* 36, no. 4, pp. 428–436.

Climate: Blame it on El Niño. (June 1990). *USA Today* (Newsletter Edition)—*The World of Science* 118, pp. 1–2.
Studying El Niño may give insight to possible repercussions of global warming.

Cook, E. R. (1993). Using tree rings to study past El Niño/Southern Oscillation influences on climate. In H. F Diaz and V. Markgraf, eds., *El Niño: Historical and Paleoclimatic Aspects of the Southern Oscillation*, pp. 203–214.

Cooke, R. (January 18, 1994). El Niño has a new found sister. *Newsday*, "Discovery," p. 57.
Discussion of La Niña, the cold phase of the Southern Oscillation.

Current causes climatic chaos. (August 1983). *Science Digest* 91, p. 27.
Canadians give an early warning of El Niño's complicating role.

D'Arrigo, R. D., and Jacoby, G. C. (1991). A 1000-year record of winter precipitation from northwestern New Mexico, USA: a reconstruction from tree-rings and its relation to El Niño and the Southern Oscillation. *The Holocene: An Interdisciplinary Journal* 1, no. 2, p. 95.

D'Arrigo, R. D., and Jacoby, G. C. (1993). A tree-ring reconstruction of New Mexico winter precipitation and its relation to El Niño/Southern Oscillation events.: In H. P. Diaz and V. Markgraf, eds., *El Niño: Historical and Paleoclimatic Aspects of the Southern Oscillation*, pp. 243–257.

Delworth, T. L., Manabe, S., and Stouffer, R. J. (1993). Interdecadal variations of the thermohaline circulation in a coupled ocean-atmosphere model. *Journal of Climate* 6, pp. 1993–2011.

Deser, C. (1993). Blackmon: Surface climate variations over the North Atlantic Ocean during winter: 1900–1989. *Journal of Climate* 6, pp. 1743–1753.

Deser, C., and Wallace, J. M. (1987). El Niño events and their relation to the Southern Oscillation: 1925–86. *Journal of Geophysical Research* 92, no. C13, pp. 14189–14196.

DeVries, T. J. (1987). A review of geological evidence for ancient El Niño activity in Peru. *Journal of Geophysical Research* 92, no. C13, pp. 14471–14479.

Diaz, H. F., and Markgraf, V. (1993). El Niño: Historical and Paleoclimatic Aspects of the Southern Oscillation. Cambridge: Cambridge University Press, 476 pp.

Diaz, H. F., and Pulwarty, R. S. (1993). A comparison of Southern Oscillation and El Niño signals in the tropics. In H. F. Diaz and V. Markgraf, eds., *El Niño: Historical and Paleoclimatic Aspects of the Southern Oscillation.* pp. 175–192.

Donguy, J. R. (1987). Recent advances in the knowledge of the climatic variations in the Tropical Pacific Ocean. *Progress in Oceanography* 19, no. 1, pp. 49–85.
An area particularly sensitive to interannual oscillations associated with El Niño.

Dovgalyuk, V. V., and Klimenko, V. V. (1996). On long-term variations in the intensity of El Niño occurrences. *Geophysical Research Letters* 23, no. 25, pp. 3771–3774.

Effects of El Niño to continue until June, experts say. (March 31, 1998). *Baltimore Sun*, p. 7A.

El Niño and changing climate. (1997). Landover, MD: Federal Document Clearing House. 120 min. Videocassette.

Elsner, J. B., and Tsonis, A. A. (1991). Do bidecadal oscillations exist in the global temperature record? *Nature* 353, no. 6344, pp. 551–553.

Enfield, D. B. (1987). Progress in understanding El Niño. *Endeavour, New Series* 11, no. 4, pp. 197–204.

Enfield, D. B. (1993). Historical and prehistorical overview of El Niño/ Southern Oscillation. In H. F. Diaz and V. Markgraf, eds., *El Niño: Historical and Paleoclimatic Aspects of the Southern Oscillation*, pp. 95–117.

Finney, B. R. (1985). Anomalous westerlies, El Niño, and the colonization of Polynesia. *American Anthropologist* 87, no. 1, pp. 9–26.

FitzGerald, L. M. (October 1991). Earth keeps its cool. *Sea Frontiers* 37, pp. 12–13. Could El Niño help moderate global warming?

Flohn, H. (1986). Singular events and catastrophes now and in climatic history. *Die Naturwissenschaften* 73, no. 3, pp. 136–49.
Describes an extreme singular event of global extension such as the Super El Niño of 1982–1983.

Gates, D. M. (1993). *Climate Change and Its Biological Consequences.* Sunderland, MA: Sinauer Associates.

Gold, S. D. (2000). *Blame It on El Niño.* Austin, TX: Raintree Steck-Vaughn.

Golden, F. (April 11, 1983). Tracking that crazy weather. *Time* 121, p. 67.

Golnaraghi, Maryam, and Kaul Rajiv. (January/February 1995). The science of policy making: responding to El Niño. *Environment* 37, pp. 16–44.

Goodman, S. J., Buechler, D. E., and McCaul, E. W., Jr. (1987). (The 1997–1998 El Niño event and related wintertime lightning variations in the southeastern United States. *Geophysical Research Letters* 27, no. 4, pp. 245–246.

Gopinathan, C. K., and Sastry, J. S. (1990). Relationship between Indian summer monsoon rainfall and position of Pacific Ocean warm pool. *Indian Journal of Marine Sciences* 19, no. 4, pp. 246–250.

Green, L., and Stephanie R. B. (January 1984). El Niño: the temperamental child. *Outdoor Life* 173, pp. 32–33, 92–94.

Grimm, A. M., Barros, V. R., and Doyle, M. E. (2000). Climate variability in southern South America associated with El Niño and La Niña events. *Journal of Climate* 13, no. 1, pp. 35–58.

Gunn, J. (1991). Influences of various forcing variables on global energy balance during the period of intensive instrumental observation (1958–1987) and their implications for paleoclimate. *Climatic Change* 19, no. 4, pp. 393–420.
The ENSO appears to act as a hemispheric energy balancing mechanism.

Gutzler, D. S., and Rosen, R. D. (1992). Interannual variability of wintertime snow cover across the Northern Hemisphere. *Journal of Climate* 5, pp. 1441–1447.

Hansen, D. V. (1990). Physical aspects of the El Niño event of 1982–1983. In P. W. Glynn, ed., *Ecological Consequences of the 1982–83 El Niño-Southern Oscillation*, pp. 1–20.

Hansen, D. V., and Bezdek, H. F. (1996). On the nature of decadal anomalies in North Atlantic Sea Surface Temperature. *Journal of Geophysical Research* 101, pp. 9749–9758.

Hsieh, W., D. Ware, M., and Thomson, R. E. (1995). Wind induced upwelling along the west coast of North America, 1899–1998. *Canadian Journal of Fisheries & Aquatic Sciences* 52, pp. 325–334.

Hurrell, J. W. (1995). Decadal trends in the North Atlantic Oscillation: regional temperatures and precipitation. *Science* 269, pp. 676–679.

Inclement weather? Blame it on El Niño (Spring 1983). *Earth Science* 36, pp. 7–8.

Ismail, S. A. (1987). Long-range seasonal rainfall forecast for Zimbabwe and its relation with El Niño/Southern Oscillation (ENSO). *Theoretical and Applied Climatology* 38, no. 2, pp. 93–102.

Jacobs, G. A., Hurlburt, H. E., Kindle, J. C., Metzger, E. J.; Mitchell, J. L., Teague, W. J., and Wallcraft, A. J. (1994). Decade-scale trans-Pacific propagation and warming effects of an El Niño anomaly. *Nature* 370, no. 6488, pp. 360–363.

Juillet-Leclerc, A., Labeyrie, L. D.; Reyss, J. L., and Schrader, H. (1991). Temperature variability in the Gulf of California during the last century: a record of the recent strong El Niño. *Geophysical Research Letters* 18, no. 10, pp. 1889–1892.

Kahya, E., and Dracup, J. A. (1994). The influences of Type 1 El Niño and

La Niña events on streamflows in the Pacific Southwest of the United States. *Journal of Climate* 7, no. 6, pp. 965–976.

Kane, R. P. (1992). Relationship between QBOs of stratospheric winds, ENSO variability and other atmospheric parameters. *International Journal of Climatology* 12, no. 5, pp. 435–447.

Kane, R. P. (1999). Some characteristics and precipitation effects of the El Niño of 1997–1998. *Journal of Atmospheric and Solar-Terrestrial Physics* 61, no. 18, pp. 1325–1346.

Kane, R. P. (2000). El Niño/La Niña relationship with rainfall at Huancayo, in the Peruvian Andes. *International Journal of Climatology* 20, no. 1, pp. 63–72.

Kerr, R. A. (February 13, 1987). Another El Niño surprise in the Pacific, but was it expected? *Science* 235, pp. 744–745.

Kerr, R. A. (May 13, 1988). Weather in the wake of El Niño. *Science* 240, p. 883.
Looking back, and interpreting, the 1986–1987 event.

Kok, C. J., and Opsteegh, J. D. (1985). Possible causes of anomalies in seasonal mean circulation patterns during the 1982–83 El Niño event. *Journal of the Atmospheric Sciences* 42, no. 7, pp. 677–694.

Li, Z., and Kafatos, M. (2000). Interannual variability of vegetation in the United States and its relation to El Niño/Southern Oscillation. *Remote Sensing of the Environment* 71, no. 3, pp. 239–247.

Liu, Z., J., Kutzbach, J., and Wu, L. (2000). Modeling climate shift of El Niño variability in the Holocene. *Geophysical Research Letters* 27, no. 15, pp. 2265–2268.

Lough, J. M. (1994). Climate variation and El Niño–Southern Oscillation events in the Great Barrier Reef: 1958 to 1987. *Coral Reefs* 13, no. 3, pp. 181–195.

Lough, J. M., and Fritts, H. C. (1990). Historical aspects of El Niño/Southern Oscillation—information from tree rings. In P. W. Glynn, ed., *Global Ecological Consequences of the 1982–83 El Niño–Southern Oscillation*, pp. 285–321.

Lowenstein, F. (November 1986). Trends: Listening to the weather. *Technology Review* 89, pp. 8–9.

Martin, L., Absy., M. L., Flexor, J.-M., Fournier, M., Mourguiart, P., Sifeddine, A., and Turcq, B. (1992). Records of El Niño-like conditions in South America during the last 7000 years. *Comptes Rendus de l'Academie des Sciences, Serie II (Mecanique, Physique, Chimie Sciences de la Terre et de l'Univers)* 315, no. 1, pp. 97–102.

McGowan, J. A. (1984). The California El Niño, 1983. *Oceanus* 27, no. 2, pp. 48–51.

McPhaden, M. J. (August 4, 1994). The eleven-year El Niño? *Nature* 370, p. 326.

Moseley, M. E., and Richardson, J. B. (November/December 1992). Doomed by natural disaster. *Archeology* 45, pp. 44–45.

Moses, T., Kiladis, G. N., Diaz, H. F., and Barry, R. G. (1987). Character-istics and frequency reversals in mean sea level pressure in the North Atlantic sector and their relationships to long-term temperature trends. *Journal of Climatology* 7, pp. 13–30.

Nash, J. M. (August 18, 1997). Is it El Niño of the century? *Time* 150, pp. 56–58.

Nash, M. (February 16, 1998). The fury of El Niño. *Time.*

Oliwenstein, L. (August 1989). Ghosts of Christmas. *Discover* 22, p. 10.
A bird's eye view of El Niño.

Ortlieb, L., and Machare, J. (1993). Former El Niño events: records from western South America. *Global & Planetary Change* 7, no. 1–3, pp. 181–202.

Pariwono, J. I., Bye, J.A.T., and Lennon, G. W. (1986). Long-period var-iations of sea-level in Australasia. *Geophysical Journal of the Royal Astronomical Society* 87, no. 1, pp. 43–54.
These features are related to the ENSO cycle which for the first time is linked, inter alia, with Southern Ocean mechanisms.

Parker, D. E., and Folland, C. K. (1988). The nature of climatic variability. *Meteorology Magazine* 117, pp. 201–210.

Philander, S.G.H. (1993). Anomalous El Niño of 1982–83. *Nature* 305, no. 5929, p. 16.

Phillips, D. (January/February 1993). Blame it on El Niño. *Canadian Ge-ographic* 113, pp. 20–21.

Picaut, J., and Delcroix, T. (1995). Equatorial wave sequence associated with warm pool displacements during the 1986–1989 El Niño-La Niña. *Journal of Geophysical Research* 100, no. C9, pp. 18393–18408.

Polonsky, A. B. (1994). Comparative study of the Pacific ENSO event of 1991–92 and the Atlantic ENSO-like event of 1991. *Australian Journal of Marine & Freshwater Research* 45, no. 4, pp. 705–725.

Ponte, R. M. (1986). The statistics of extremes, with application to El Niño. *Reviews of Geophysics* 24, no. 2, pp. 285–297.

Prasad, K. D., and Singh, S. V. (1996). Seasonal variations of the relation-ship between some Enso parameters and Indian rainfall. *International Journal of Climatology* 16, no. 8, pp. 923–933.

Quinn, W. H., and Neal, V. T. (1983). Long-term variations in the South-ern Oscillation, El Niño, and Chilean subtropical rainfall. *Fishery Bul-letin* 81, no. 2, pp. 363–374.

Quinn, W. H., and Neal, V. T. (1987). Antunez de Mayolo, S. E.: El Niño occurrences over the past four and a half centuries. *Journal of Geo-physical Research* 92, no. C13, pp. 14449–14461.

Quinn, W. H., Zopf, David O., Short, Kent S., Kuo Yang, and Richard T. W. (1981). Historical trends and statistics of the Southern Oscillation, El Niño, and Indonesian droughts (Peru). *Fishery Bulletin* 76, no. 3, pp. 663–678.

Quiroz, R. S. (1983). The climate of the "El Niño" winter of 1982/83— a season of extraordinary climatic anomalies. *Monthly Weather Review* 111, no. 8, pp. 1685–1706.

Ramage, C. S. (1975). Preliminary discussion of the meteorology of the 1972–73 El Niño. *Bulletin of the American Meteorological Society* 56, no. 2, pp. 234–242.

Rasmusson, E. M. (1985). El Niño and variations in climate. *American Scientist* 73, no. 2, pp. 168–177.

Rasmusson, E. M., and Carpenter, T. H. (1983). The relationship between eastern equatorial Pacific sea surface temperatures and rainfall over India and Sri Lanka. *Monthly Weather Review* 111, no. 3, pp. 517–528.

Monsoon season (June–September) precipitation data from 31 Indian subdivisions and mean monthly precipitation data from 35 Indian and Sri Lanka stations, spanning the period 1875–1979, were analyzed to determine the relationship between equatorial Pacific warm episodes (El Niño events) and interannual fluctuations in precipitation over India and Sri Lanka.

Rasmusson, E. M., and Hall, J. M. (August 1983). El Niño: The great Equatorial Pacific Ocean warming event of 1982–1983. *Weatherwise* 36, pp. 166–175. A concise, readable report on this occurrence of El Niño.

Reiter, E. R. (1978). The interannual variability of the ocean–atmosphere system. *Journal of the Atmospheric Sciences* 35, no. 3, pp. 349–370. The trade wind surges also are related to El Niño through a feedback involving the hydrological cycle and upwelling of cold water forced by Ekman pumping.

Rippey, B. (May 3, 1994). A tale of three Niños: 1911–1913, 1939–42, 1991–93. *Weekly Weather and Crop Bulletin* 81, pp. 10–11.

Rittenour, T. M., Brigham-Grette, J., and Mann, M. E. (2000). El Niño-like climate teleconnections in New England during the late Pleistocene. *Science* 288, no. 5468, pp. 1039–1042.

Rogers, J. C. (1990). Patterns of low-frequency monthly sea level pressure variability (1899–1986) and associated wave cyclone frequencies. *Journal of Climate* 3, pp. 1364–1379.

Rong-Hua Zhang, and Endoh, M. (1994). Simulation of the 1986–1987 El Niño and 1988 La Niña events with a free surface tropical Pacific Ocean general circulation model. *Journal of Geophysical Research* 99, no. C4, pp. 7743–7759.

Sandweiss, D. H., Richardson, J. B., III, Reitz, E. J., Rollins, H. B., and Maasch, K. A. (1996). Geoarchaeological evidence from Peru for a

5000 years B. P. onset of El Niño. *Science* 273, no. 5281, pp. 1531–1533.

Schell, I. I. (1965). The origin and possible prediction of the fluctuations in the Peru Current and upwelling. *Journal of Geophysical Research* 70, pp. 5529–5540.

Schonwiese, C. D., and Birrong, W. (1990). European precipitation trend statistics 1851–1980 including multivariate assessments of the anthropogenic CO/sub 2/signal. *Zeitschrift fur Meteorologie* 40, no. 2, pp. 92–98.
A multivariate regression model including volcanic and solar forcing, ENSO is used.

Serreze, M., Carse, F., and Barry, R. G. (1997). Icelandic low cyclone activity: climatological features, linkages with the NAO, and relationship with recent changes in the Northern Hemisphere circulation. *Journal of Climate* 10, pp. 453–464.

Simpson, J. J. (1994). A simple model of the 1982–83 California "El Niño" *Geophysical Research Letters* 11, no. 3, pp. 237–240.

Soreide, N., and McPhaden, M. (1995). El Niño watching on World Wide Web. *Sea Frontiers* 41, pp. 26–29.

Stevens, W. K. (June 21, 1997). El Niño is back, scientist says, with threat of global havoc. *New York Times*, p. 7.

Stevens, W. K. (August 26, 1997). Forecasters begin to see who will feel effects of El Niño. *New York Times*, C4.

Stockton, C. W., and Glueck, M. F. (January 1999). Long-term variability of the North Atlantic oscillation (NAO). Proceedings of the American Meteorological Society Tenth Symposium on Global Change Studies, Dallas, TX, pp. 290–293.

TOGA COARE: Unlocking the mystery of El Niño. (1993). Produced by John Kermond. NOAA Office of Global Programs. 21 min. Videocassette.

Trenberth, K. E. (1990). Recent observed interdecadal climate changes in the Northern Hemisphere, *Bulletin of the American Meteorological Society* 71, no. 7, pp. 988–993.
North Pacific changes appear to be linked through teleconnections to tropical atmosphere–ocean interactions and the frequency of El Niño.

Trenberth, K. E., and Hoar, T. J. (1996). The 1990–1995 El Niño-Southern Oscillation event: longest on record. *Geophysical Research Letters* 23, no. 1, pp. 57–60.

Trenberth, K. E., and Hurrell, J. W. (1999). Comment on the interpretation of short climate records with comments on the North Atlantic and Southern Oscillations. *Bulletin of the American Meteorological Society*.

Van Der Hammen, T., and Cleef, A. M. (1992). Holocene changes of rainfall and river discharge in northern South America and the El Niño phenomenon. *Erdkunde* 46, no. 3–4, pp. 252–256.

Verma, R. K. (1990). Recent monsoon variability in the global climate perspective. *Mausam* 41, no. 2, pp. 315–320.

Vogel, S. (July 1989). Deep-rooted disturbance. *Discover* 10, pp. 26–29. Do earthquakes at the boundary of tectonic plates help drive El Niño?

Wang Shaowu. (1992). Reconstruction of El Niño event chronology for the last 600 year period. *Acta Meteorologica Sinica* 6, no. 1, pp. 47–57.

Weather maps circa 2000 BC. (October 23, 1993). *Science News* 144, p. 270. Finding evidence for historic El Niños.

Weber, G.-R. (1990). North Pacific circulation anomalies, El Niño and anomalous warmth over the North American continent in 1986–1988: possible causes of the 1988 North American drought. *International Journal of Climatology* 10, no. 3, pp. 279–289.

Webster, P. J., and Palmer, T. N. (December 11, 1997). The past and future of El Niño. *Nature* 390, pp. 562–564.

Wells, L. E. (1987). An alluvial record of El Niño events from northern coastal Peru. *Journal of Geophysical Research* 92, no. C13, pp. 14463–14470.

Wells, L. E. (1990). Holocene history of the El Niño phenomenon as recorded in flood sediments of northern coastal Peru. *Geology* 18, no. 11, pp. 1134–1137.

What's next for El Niño? (December 1983). *Sunset Magazine* 171, p. 224.

Whetton, P. H., and Rutherfurd, I. (1994). Historical ENSO teleconnections in the eastern hemisphere. *Climatic Change* 28, no. 3, pp. 221–253.

Wilby, R. (1993). Evidence of ENSO in the synoptic climate of the British Isles since 1880. *Weather* 48, no. 8, pp. 234–239.

Wolter, K., and Timlin, M. (September 1998). Measuring the strength of ENSO events, how does 1997/98 rank? *Weather* 3, no. 9, pp. 315–324.

Wolter, K., and Timlin, M. S. Monitoring ENSO in COADS with a seasonally adjusted principle component index. Proceedings of the 17th Climate Diagnostics Workshop, Norman, OK, pp. 52–57.

Wunsch, C. (1999). The interpretation of short climate records, with comments on the North Atlantic and Southern Oscillations. *Bulletin of the American Meteorological Society* 80, p. 257.

Zhan Shuyun, and Yang Shurui. (1990). An analysis of El Niño processes during 1949–1987. *Acta Oceanologica Sinica* 9, no. 2, pp. 219–226.

Index

About the Authors

JOSEPH S. D'ALEO has 30 years experience in professional meteorology. He has BS and MS degrees in meteorology from the University of Wisconsin and did doctoral studies in meteorology at New York University. He taught meteorology at the college level for over eight years and was a cofounder and the first director of meteorology at the cable TV Weather Channel, a position he held for seven years. He joined Weather Services International (WSI) in 1989, where he was a marketing manager and chief meteorologist. Currently, he is senior editor (also known as "Dr. Dewpoint") for WSI's Intellicast Web site. D'Aleo is a Certified Consultant Meteorologist and was elected a Fellow of the American Meteorological Society. He has authored, presented, and published numerous papers focused on advanced applications enabled by new technologies and how research into ENSO and other atmospheric and oceanic phenomena has made possible skillful seasonal forecasts.

PAMELA G. GRUBE has 27 years of experience in the field of meteorology. She has a BA in Earth and Space Science from Millersville University in Pennsylvania, and an MS in Atmospheric Science from Colorado State University. She did her Master's thesis work under tropical expert Dr. William Gray, analyzing data from the tropical Atlantic. Her experience ranges from weather modification research at the University of Wyoming to 12 years of operational forecasting at The Weather Channel, to her current position as Assistant Professor of Meteorology at Lyndon State College in Lyndonville, VT.